Maths for A Level **Chem**

A Course Companion

Updated Edition

Stephen Doyle

Supports A Level Chemistry courses from AQA, Edexcel, OCR, WJEC, CCEA, the International Baccalaureate and the Cambridge Pre-U

Published in 2016 by Illuminate Publishing Ltd, P.O Box 1160,
Cheltenham, Gloucestershire GL50 9RW

First edition published by Illuminate Publishing in 2014

Orders: Please visit www.illuminatepublishing.com
or email sales@illuminatepublishing.com

British Library Cataloguing in Publication Data

A catalogue record for this book is available from the British Library

ISBN 978-1-908682-35-2

Printed and bound in England by Cambrian Printers
01.16

The publisher's policy is to use papers that are natural, renewable and recyclable products made from wood grown in sustainable forests. The logging and manufacturing processes are expected to conform to the environmental regulations of the country of origin.

Editor: Geoff Tuttle
Cover and text design: Nigel Harriss
Text and layout: GreenGate Publishing Services, Tonbridge, Kent

Acknowledgements

I am very grateful to Peter, Geoff and the team at Illuminate Publishing for their professionalism, support and guidance throughout this project. It has been a pleasure to work so closely with them.

The publisher and author would like to thank the following for their help and advice in reviewing this book:

Peter Blake

Elizabeth Humble

Photo credits

Cover © Shutterstock – Matej Kastelic; **p.11** © Klekta Darya – Fotolia.com; **p.19** © Vladru – Shutterstock; **p.20** © Eugene Sergeev – Shutterstock; **p.21** © Iaryna Turchyniak – Shutterstock; **p.23** © Sofiaworld – Shutterstock; **p.37** © Autre – Fotolia.com; **p.42** © Chris Lenfert – Shutterstock; **p.49** © 3DStyle – Shutterstock; **p.58** © marcel – Fotolia.com; **p.66** © general-fmv – Shutterstock; **p 80** © Decha Thapanya – Shutterstock; **p.83** © sumire8 – Fotolia.com; **p.88** © Nneirda – Fotolia.com; **p.89** © airborne77 – Fotolia.com; **p.94** © Neyro – Fotolia.com; **p.117** © FMax – Fotolia.com; **p.118** © antonio scarpi – Fotolia.com; **p.119** © ollaweila – Fotolia.com; **p.122** © PRILL Mediendesign – Fotolia.com; **p.157** © perfectmatch – Fotolia.com; **p.163** © Swapan – Fotolia.com

Contents

Introduction

The aim of this book is to help you with the numerical parts of AS and A-level Chemistry if you feel unsure about mathematics. In Chapters 1–5 you will be introduced to those mathematical techniques which you will need throughout the course. It will be assumed that you have completed a GCSE Mathematics course so there will be sections with which you feel ok and don't need to cover. After spending time on this section you will feel more confident about attempting calculations and you can concentrate on the chemistry rather than the mathematical concepts involved.

In Chapters 6–27 on chemistry calculations, you will come across maths tips which will help explain some of the maths that you will come across in the questions.

How to use this book

This book is a combination of chemistry and mathematics. The first five chapters of the book cover the main mathematical concepts needed for AS and A-level Chemistry or other equivalent qualifications. Nearly all the specifications have a list of the mathematics you will need to know and the first five chapters of this book cover the main parts, with the rest of the mathematics picked up in the context of chemistry in the chemistry chapters of the book. You need to look at the first five chapters to see which areas you need to concentrate on and you can then work through these chapters to build up your skills and assess your understanding by attempting the Test yourself questions and checking the answers.

If you feel you have a reasonable understanding of the mathematics needed, then you may choose to use the mathematics section as a reference section to be used when answering chemistry questions in the later chapters.

Chapters 6 to 27 cover the chemistry content and the emphasis on this is to provide explanations where any mathematics is involved. The chemistry content covers only those aspects of chemistry where there are calculations or some other aspects of mathematics such as graphs or geometry. Although many chemistry topics are interrelated, there is a logical pathway through the material and the book follows this. You will need to look at the Specification map at the end of the book to check if you need to cover all of the chemistry content or just some of it. As the main emphasis of the book is the mathematical aspects of chemistry, the mathematical concepts are explained in more detail than in the majority of AS and A-level text books. As this is a mathematics for chemistry book, this book does not cover all the chemistry, only that with a mathematics element.

There are several ways this book will help you:

1 Chapters 1 to 5 will offer a refresher course in the mathematical knowledge and skills you will need in order to succeed at AS, A or equivalent chemistry. You should work through this section to refresh your knowledge of maths before you start the chemistry course. As this is the core mathematical knowledge needed, you should ensure you are proficient before you attempt the chemistry calculations.

2 Other mathematical concepts such as the use of logarithms are dealt with in the chapter where they are used. For example, the use of logarithms is explained in Chapter 26 on calculations involving pH as this is the only section where you will come across them.

3 Chapters contain **Pointers** in boxes which remind you of important points and offer advice on things to watch out for. There are Pointers throughout, which explain any tricky mathematics parts and also Pointers which point out common mistakes or misconceptions.

Notice the way that we write 1.0 rather than just 1. This is to show that we have written the number to one decimal place.

4 You can use the **Specification map** to check which parts of the book you need to cover for your particular specification/syllabus.

Specification map

Ticks indicate the mathematical requirements that are stated in the specifications.

Topic	Section	WJEC	AQA	OCR	Edexcel	CIE	CCEA	IB	Cam Pre-U
1 Arithmetic and numerical computation									
Order when evaluating expressions	1.1	✓	✓	✓	✓	✓	✓	✓	✓
Directed numbers	1.3	✓	✓	✓	✓	✓	✓	✓	✓
Calculations involving fractions	1.4	✓	✓	✓	✓	✓	✓	✓	✓
Percentages, percentage difference	1.5	✓	✓	✓	✓	✓	✓	✓	✓
Indices and standard form	1.6	✓	✓	✓	✓	✓	✓	✓	✓
Ratio and proportion	1.7	✓	✓	✓	✓	✓	✓	✓	✓
2 Handling data									
Approximating numbers (d.p. and s.f.)	2.1	✓	✓	✓	✓	✓	✓	✓	✓
Making estimates of the results of calculations	2.2	✓	✓	✓	✓	✓	✓	✓	✓
Finding the arithmetic mean and weighted mean	2.3	✓	✓	✓	✓	✓	✓	✓	✓
Units in chemistry	2.4	✓	✓	✓	✓	✓	✓	✓	✓
The ideal gas equation/converting volumes	2.5	✓	✓	✓	✓	✓	✓	✓	✓
3 Algebra									
Symbols used in algebra (=, <, <<, >>, >, ∝)	3.1	✓	✓	✓	✓	✓	✓	✓	✓
Transposition of formulae	3.2	✓	✓	✓	✓	✓	✓	✓	✓
Representing a variable using more than one letter	3.3	✓	✓	✓	✓	✓	✓	✓	✓
Solving equations	3.4	✓	✓	✓	✓	✓	✓	✓	✓
Logarithms	3.5	✓	✓	✓	✓	✓	✓	✓	✓

5 **Worked examples** are given throughout the text and these examples are very similar to examination questions. You need to work through these examples carefully, paying particular attention to the way the answers are set out.

6 At the end of each chapter there are **Test yourself questions** and these present an opportunity for you to try questions and check your answers with the worked solutions given. Be sure that you attempt all these questions to ensure you have understood the material in the chapters.

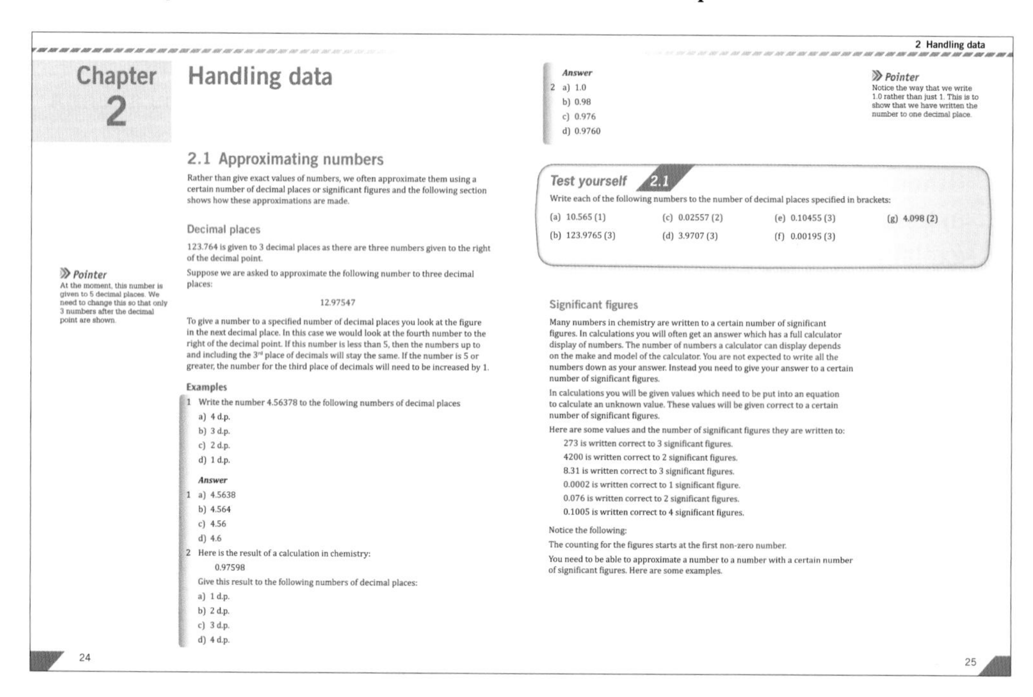

Chapter 2

Handling data

2.1 Approximating numbers

Rather than give exact values of numbers, we often approximate them using a certain number of decimal places or significant figures and the following section shows how these approximations are made.

Decimal places

123.764 is given to 3 decimal places as there are three numbers given to the right of the decimal point.

Suppose we are asked to approximate the following number to three decimal places:

12.97547

» Pointer
At the moment, this number is given to 5 decimal places. We need to change this so that only 3 numbers after the decimal point are shown.

To give a number to a specified number of decimal places you look at the figure in the next decimal place. In this case we would look at the fourth number to the right of the decimal point. If this number is less than 5, then the numbers up to and including the 3rd place of decimals will stay the same. If the number is 5 or greater, the number for the third place of decimals will need to be increased by 1.

Examples

1 Write the number 4.56378 to the following numbers of decimal places
a) 4 d.p.
b) 3 d.p.
c) 2 d.p.
d) 1 d.p.

Answer
1 a) 4.5638
b) 4.564
c) 4.56
d) 4.6

2 Here is the result of a calculation in chemistry:
0.97598
Give this result to the following numbers of decimal places:
a) 1 d.p.
b) 2 d.p.
c) 3 d.p.
d) 4 d.p.

24

2 Handling data

Answer
2 a) 1.0
b) 0.98
c) 0.976
d) 0.9760

» Pointer
Notice the way that we write 1.0 rather than just 1. This is to show that we have written the number to one decimal place.

Test yourself 2.1

Write each of the following numbers to the number of decimal places specified in brackets:

(a) 10.565 (1) (c) 0.02557 (2) (e) 0.10455 (3) (g) 4.098 (2)
(b) 123.9765 (3) (d) 3.9707 (3) (f) 0.00195 (3)

Significant figures

Many numbers in chemistry are written to a certain number of significant figures. In calculations you will often get an answer which has a full calculator display of numbers. The number of numbers a calculator can display depends on the make and model of the calculator. You are not expected to write all the numbers down as your answer. Instead you need to give your answer to a certain number of significant figures.

In calculations you will be given values which need to be put into an equation to calculate an unknown value. These values will be given correct to a certain number of significant figures.

Here are some values and the number of significant figures they are written to:

273 is written correct to 3 significant figures.
4200 is written correct to 2 significant figures.
8.31 is written correct to 3 significant figures.
0.0002 is written correct to 1 significant figure.
0.076 is written correct to 2 significant figures.
0.1005 is written correct to 4 significant figures.

Notice the following:

The counting for the figures starts at the first non-zero number.

You need to be able to approximate a number to a number with a certain number of significant figures. Here are some examples.

25

Chapter 1 Arithmetic and numerical computation

In this section you will be looking at some basic arithmetic such as the order when evaluating expressions, directed numbers, fractions, percentages, etc.

1.1 The order when evaluating an expression

In chemistry, you will often have to evaluate an expression. This involves following a series of calculations in order to arrive at a final answer. You will remember from your GCSE Maths course that there is an order to be used when evaluating expressions. The word BODMAS is often used to help remember the order in which the calculations should be carried out. BODMAS stands for Brackets, Orders, Division, Multiplication, Addition and Subtraction. Any expressions should be worked out in the order (i.e. BODMAS) starting with brackets, then orders (i.e. powers and roots), then division, then multiplication, then addition and finally subtraction.

Examples

1 Evaluate $2 + 3 \times 4$

$2 + 3 \times 4 = 2 + 12 = 14$

2 Evaluate $5 \times 8 - 10 \div 2$

According to BODMAS the division is performed and then the multiplication and finally the subtraction.

Hence, $5 \times 8 - 10 \div 2 = 40 - 5 = 35$

3 Evaluate $3 \times (24 - 9)$

According to BODMAS, the expression in the brackets is evaluated first

$3 \times (24 - 9) = 3 \times 15 = 45$

4 Evaluate the expression

$172 - 8 \times 3^2$

There are no brackets to deal with so orders (i.e. powers and roots) are dealt with first.

Hence, $172 - 8 \times 3^2 = 172 - 8 \times 9 = 172 - 72 = 100$

5 Evaluate $\frac{21 + 7}{4}$

Here the division line acts like brackets so the addition can be considered to be in brackets and should therefore be performed first.

$$\frac{21 + 7}{4} = \frac{28}{4} = 7$$

» Pointer

If you perform the calculation from left to right, you will get the answer wrong. You must remember to use BODMAS. This means that the multiplication must be performed before the addition.

1.2 Adding brackets in calculations for emphasis

Sometimes brackets are added for emphasis. For example, these two expressions are identical:

$$16 \times 3 - 24 \div 6$$

$$(16 \times 3) - (24 \div 6)$$

The expression with the brackets is sometimes used to add emphasis to the fact that the division and multiplication should be performed before the subtraction.

Test yourself 1.1

Evaluate the following expressions:

a) $3 + 14 \div 2$

b) $6 \times 3 - 4 \div 2$

c) $(5 \times 12) - 15$

d) $\frac{(15 + 5)}{2}$

e) $18 - 4 \times 3$

f) $23.6 \times 3.1 - 9.8$

g) $(3 + 4)^2$

h) $4 + 5 \times 10^2$

i) $\frac{5 \times 12}{30 \div 5}$

j) $7 - \frac{18}{3}$

k) $\frac{5}{(10 + 15)}$

Test yourself 1.2

Evaluate the following expressions:

a) $\left(\frac{28}{100} \times 15\right) + \left(\frac{72}{100} \times 16\right)$

b) $\left(\frac{65}{100} \times 39\right) + \left(\frac{15}{100} \times 40\right) + \left(\frac{20}{100} \times 41\right)$

c) $\left(\frac{18.6}{100} \times 63\right) + \left(\frac{12.4}{100} \times 64\right) + \left(\frac{60}{100} \times 65\right)$

1.3 Directed numbers

Numbers can be positive or negative or zero. Numbers with a sign (i.e. positive or negative numbers) are called directed numbers. So, with numbers, we have a size and a sign (i.e. + or –). Directed numbers can go above and below a zero value. You can see this with temperature which can be positive, negative or zero. In chemistry we also use directed numbers when dealing with charges on ions when atoms lose or gain electrons.

Thinking of directed numbers in terms of temperature

A good way to deal with directed numbers is to think of a thermometer. A thermometer has positive values above zero and negative values below zero.

Suppose you want to work out −3 + 6 you would start at the first number (i.e. −3) and imagine it on a thermometer. The next number (i.e. +6) means that you will go up 6 units. The first 3 units of the 6 will take you to zero and the next three units will take you three units above zero to give the answer +3 or just 3. Notice that we don't usually add the sign to the answer unless it is negative.

To work out −1 − 4 you would start at −1 and then go down 4 units (because the sign in front of the 4 is negative). You would then arrive at the answer −5.

Calculations involving more than two directed numbers

Suppose you have to work out the value (i.e. evaluate) −5 + 4 − 3

Find out the result of the first two numbers i.e. −5 + 4 = −1

and then use the result with the final number like this −1 − 3 = −4

Test yourself 1.3

Evaluate each of the following:

a) −5 + 4	c) 8 − 9	e) −1 − 4	g) 0 − 3	i) −2 + 1
b) −3 − 3	d) 6 − 3	f) −3 + 4	h) −7 + 5	j) − 3 − 8

Test yourself 1.4

Evaluate each of the following:

a) − 5 + 4 − 3	c) 3 − 5 + 4	e) −5 + 4 − 4	g) −3 − 5 + 4
b) −1 + 6 − 5	d) −3 + 1 + 3	f) 2 − 6 + 2	h) −1 + 7 − 4

Multiplication and division of directed numbers

The rules for the multiplication of directed numbers are that if the two numbers being multiplied are both the same sign (i.e. both positive or both negative), then the answer will be positive. If the two numbers being multiplied have opposite signs, then the answer will be negative.

These rules are summarised here:

+ × + = +
− × − = +
+ × − = −
− × + = −

The rules determining the sign of two directed numbers being divided are the same as those for multiplication. These are summarised here:

+ ÷ + = +
− ÷ − = +
+ ÷ − = −
− ÷ + = −

Test yourself 1.5

Work out the following calculations:

a) -3×-2
b) 4×-1
c) -3×1
d) 6×-2
e) $3 \div -1$
f) $10 \div -2$
g) $-12 \div -4$
h) $6 \div -3$
i) $4 \div -1$
j) $(-3)^2$
k) $-8 \div -4$
l) $(-3) \times (-6)$

1.4 Calculations involving fractions

It is assumed here that you are able to add and subtract fractions. Addition and subtraction of fractions is rarely needed in chemistry, so here we will be concentrating on multiplication and division.

Multiplication of fractions

Suppose you want to multiply these two fractions:

$$\frac{2}{3} \times \frac{5}{8}$$

Look at the diagonal containing the numbers 5 and 3 and ask is there a number which can be divided into both numbers. There is only 1 that divides exactly into 5 and 3. We now look at the diagonal containing the numbers 2 and 8. As two divides exactly into 2 and 8, we do this and obtain the following

$$\frac{2}{3} \times \frac{5}{8} = \frac{1}{3} \times \frac{5}{4}$$

Now we multiply the top numbers together and then the bottom numbers together to give

$$\frac{1}{3} \times \frac{5}{4} = \frac{5}{12}$$

Just check that there is not a number that divides into both the top and the bottom. There isn't in this case, so the final answer is $\frac{5}{12}$.

Examples

1 $\frac{15}{28} \times \frac{7}{20} = \frac{3}{4} \times \frac{1}{4}$

$= \frac{3}{16}$

2 $\frac{3}{4} \times \frac{2}{15} = \frac{1}{2} \times \frac{1}{5} = \frac{1}{10}$

3 $\frac{5}{14} \times 64 = \frac{5}{14} \times \frac{64}{1} = \frac{5}{7} \times \frac{32}{1} = \frac{160}{7} = 22\frac{6}{7}$

Pointer
15 and 20 can both be divided by 5 and 28 and 7 can both be divided by 7.

Pointer
Always check that this fraction cannot be cancelled further.

Division of fractions

To divide two fractions like this $\frac{4}{5} \div \frac{2}{5}$ you turn the second fraction upside down and replace the division sign with a multiplication sign.

Hence, $\frac{4}{5} \div \frac{2}{5} = \frac{4}{5} \times \frac{5}{2} = 2$

The examples here show some typical calculations you could be asked to carry out in your Chemistry course.

Examples

1 $\frac{\frac{3}{25}}{\frac{9}{50}} = \frac{3}{25} \div \frac{9}{50} = \frac{3}{25} \times \frac{50}{9} = \frac{1}{1} \times \frac{2}{3} = \frac{2}{3}$

2 $25 \div \frac{5}{16} = 25 \times \frac{16}{5} = 5 \times 16 = 80$

3 $\frac{55}{\frac{11}{25}} = 55 \times \frac{25}{11} = 5 \times 25 = 125$

4 $\frac{\frac{3}{8}}{2} = \frac{\frac{3}{8}}{\frac{2}{1}} = \frac{3}{8} \times \frac{1}{2} = \frac{3}{16}$

5 $\frac{\frac{2}{5}}{4^2} = \frac{\frac{2}{5}}{16} = \frac{\frac{2}{5}}{\frac{16}{1}} = \frac{2}{5} \times \frac{1}{16} = \frac{1}{40}$

Test yourself 1.6

Work out the following divisions giving your answers as fractions in their simplest form:

a) $\frac{\frac{3}{10}}{\frac{9}{25}}$ b) $16 \div \frac{4}{5}$ c) $\frac{24}{\frac{8}{15}}$ d) $\frac{\frac{4}{5}}{8}$ e) $\frac{\frac{3}{25}}{3^2}$

1.5 Percentages

Expressing one quantity as a percentage of another

This is done by first expressing one quantity as a fraction of another and then converting the fraction to a percentage by multiplying by 100.

Suppose a quantity A is being expressed as a percentage of a quantity B then we would write it as follows:

$$\frac{A}{B} \times 100\%$$

Example

The theoretical yield of a certain product in a reaction is 0.458 g. In an experiment the actual yield was 0.412 g. Find the actual yield as a percentage of the theoretical yield.

Note that the actual yield needs to be the top part of the fraction (i.e. the numerator) and the theoretical yield needs to be the bottom part (i.e. the denominator). The fraction then needs to be multiplied by 100 to convert it to a percentage.

Hence we have $= \frac{\text{actual yield}}{\text{theoretical yield}} \times 100 = \frac{0.412}{0.458} \times 100$

$= 89.9563\%$

$= 90.0\%$

Pointer

Values are to 3 s.f. so the final answer is given to 3 s.f. Notice that we write 90.0 rather than just 90 to show we have considered 3 s.f. rather than just 2 s.f.

Percentage difference

You will come across percentage differences when a value obtained by experiment is compared with a more accurate value (maybe obtained by a more accurate experiment or a theoretical value). The percentage difference is calculated by finding the difference in the two values and then dividing them by the more accurate value and then multiplying by 100.

$$\text{Percentage difference} = \frac{\text{Difference in the two values}}{\text{More accurate value}} \times 100$$

Example

A certain chemical reaction is exothermic (i.e. gives out heat). A student performs an experiment to measure the molar enthalpy change for this reaction and records a value of $-65.7 \text{ kJ mol}^{-1}$. When the student looked up the value in the data book, it was $-80.9 \text{ kJ mol}^{-1}$. Express the difference between the student's value and the data book value as a percentage of the data book value.

Answer

The difference between the two values $= 80.9 - 65.7$

$= 15.2 \text{ kJ mol}^{-1}$

Percentage difference $= \frac{\text{difference in two values}}{\text{data book value}} \times 100$

$= \frac{15.2}{80.9} \times 100$

$= 18.7886\%$

$= 18.8\%$ (correct to 3 s.f.)

Pointer

Do not worry about the minus signs in the values. Just remove the minus signs and subtract the smaller from the larger value.

Pointer

Always look at the calculator value and decide how many significant figures to give your answer to. In the question the values were to three significant figures so the answer should be given to this number of significant figures. If the question specifically asks for the answer to a certain number of significant figures, then you need to give your answer to this number.

1.6 Indices and standard form

The rules of indices

An index is a power. For example, in the number 10^3 the 3 is the index. The plural of index is indices and in this section you will revise the manipulation of indices.

Multiplication of indices

The general rule for multiplication of indices is:

$$x^a \times x^b = x^{a+b}$$

Pointer
For multiplication, you add the indices together.

Examples

1 $2^3 \times 2^5 = 2^{3+5} = 2^8$

2 $3^2 \times 3^4 \times 3 = 3^{2+4+1} = 3^7$

3 $10^3 \times 10^2 \times 10 = 10^{3+2+1} = 10^6$

4 $10^3 \times 10^{-1} = 10^{3+(-1)} = 10^2$

5 $10^{-3} \times 10^{-2} = 10^{-3+(-2)} = 10^{-5}$

Pointer
Note that $3 = 3^1$

Pointer
Note that
$-3 + (-2) = -3 - 2 = -5$

Division of indices

The general rule for division of indices is:

$$x^a \div x^b = x^{a-b}$$

or

$$\frac{x^a}{x^b} = x^{a-b}$$

Pointer
For division, you subtract the indices.

Examples

1 $2^6 \div 2^4 = 2^{6-4} = 2^2$

2 $\frac{3^7}{3^2} = 3^{7-2} = 3^5$

3 $\frac{10^6}{10^{-2}} = 10^{6-(-2)} = 10^8$

4 $\frac{10^{-3}}{10^{-4}} = 10^{-3-(-4)} = 10^{-3+4} = 10$

5 $\frac{10^{-5}}{10^3} = 10^{-5-3} = 10^{-8}$

Indices raised to a power

The general rule for indices raised to a power is:

$$(x^a)^b = x^{a \times b}$$

Pointer
For indices raised to a power, you multiply the index by the power.

Examples

1 $(2^4)^2 = 2^{4 \times 2} = 2^8$

2 $(10^2)^3 = 10^{2 \times 3} = 10^6$

3 $(10^{-2})^2 = 10^{(-2) \times 2} = 10^{-4}$

4 $(10^{-6})^2 = 10^{(-6) \times 2} = 10^{-12}$

5 $(10^{-5})^3 = 10^{(-5) \times 3} = 10^{-15}$

Calculations involving multiplication and division

When performing calculations involving indices, it is easiest to do the multiplication first and then the division.

Examples

1 $\dfrac{10^4 \times 10^3}{10^2 \times 10^2} = \dfrac{10^{4+3}}{10^{2+2}} = \dfrac{10^7}{10^4} = 10^{7-4} = 10^3$

2 $\dfrac{10^6 \times 10^{-3}}{10^2} = \dfrac{10^{6+(-3)}}{10^2} = \dfrac{10^3}{10^2} = 10$

3 $\dfrac{10^{-1} \times 10^6}{10^9} = \dfrac{10^{-1+6}}{10^9} = \dfrac{10^5}{10^9} = 10^{5-9} = 10^{-4}$

4 $\dfrac{10^{-6} \times 10^{-5}}{10^{-3} \times 10^{-2}} = \dfrac{10^{-6+(-5)}}{10^{-3+(-2)}} = \dfrac{10^{-11}}{10^{-5}} = 10^{-11-(-5)} = 10^{-11+5} = 10^{-6}$

5 $\dfrac{10^5 \times 10^{-10} \times 10}{10^{-5} \times 10^3} = \dfrac{10^{5+(-10)+1}}{10^{-5+3}} = \dfrac{10^{-4}}{10^{-2}} = 10^{-4-(-2)} = 10^{-4+2} = 10^{-2}$

Representing numbers in standard form

In chemistry, we often deal with very large and very small numbers and in these cases it would be impossible to represent these as ordinary numbers in calculations. Standard form is used as an alternative way of representing numbers.

Numbers in standard form are represented using the following format

$$a \times 10^n$$

where a is any number between 1 and 10 including 1 but not including 10 and n is an integer (i.e. a whole number, positive or negative including zero).

Some of the values a could take include

1, 1.2, 3.555, 9.8009

Some of the values a could not take include

0.8, 0.25, 0.999, 10, 11.4, 200

Values n could take include

1, 5, 10, 23, 0, −1, −5, −10

Values n could not take include

1.5, 0.25, −0.02, −1.5

Test yourself 1.7

Numbers in standard form are represented in the form $a \times 10^n$.

By considering the values of a and n, decide whether or not each of the following numbers is in standard form:

a) 1.00×10^3
b) 3.667×10^{-5}
c) 0.75×10^2
d) 0.8×10^{-15}
e) 1.005×10^{-8}
f) 7.5×10^{12}
g) $6 \times 10^{1.2}$
h) 5.007×10^{-25}
i) 4
j) 0.8

Converting numbers into standard form

Ordinary numbers can be converted into standard form using the following method.

Suppose you want to convert the number 3000 into standard form.

1 Decide where the decimal point would go in the number.
Here it is 3000.

2 Move the decimal point to the left so that the number remaining is a valid value of a (i.e. between 1 and 10 including 1 but not including 10) and at the same time count the number of places the decimal point has been moved.

This number is the value of n, the power to which the 10 is raised.

Hence in 3000 we move the decimal point 3 places to give 3.000, so the number in standard form is written as 3×10^3.

Examples

Convert the following numbers into standard form:

1 16
2 200
3 345
4 78 000
5 100 000
6 12.45
7 100.9
8 125.89
9 16 001
10 23.54

Note that $16 = 1.6 \times 10^1$ but since $10^1 = 10$ we write this as 1.6×10

Answers

1 $16 = 1.6 \times 10$

2 $200 = 2 \times 10^2$

3 $345 = 3.45 \times 10^2$

4 $78\,000 = 7.8 \times 10^4$

5 $100\,000 = 1 \times 10^5$

6 $12.45 = 1.245 \times 10$

7 $100.9 = 1.009 \times 10^2$

8 $125.89 = 1.2589 \times 10^2$

9 $16\,001 = 1.6001 \times 10^4$

10 $23.54 = 2.354 \times 10$

Converting numbers smaller than one into standard form

Numbers less than one such as 0.56, 0.0003 can be converted into standard form in the following way.

Move the decimal point to the right until the number remaining is a valid value for a and count the number of decimal places moved as this is the value of n. Because the number is less than one and we are moving the decimal point to the right, the value of n is made negative.

For example, converting the number 0.5, the decimal point is moved one place to the right to give 5 as the value of a. n, the number of decimal places moved is 1, hence we obtain $0.5 = 5 \times 10^{-1}$.

Converting the number 0.0003, the decimal point is moved four places to the right to give the number 3 as the value for a. n is therefore −4 so we obtain

$$0.0003 = 3 \times 10^{-4}$$

Examples

Convert the following numbers into standard form.

1 0.3

2 0.008

3 0.00105

4 0.0000357

5 0.0815

6 0.00001

7 0.004

8 0.324

9 0.000451

10 0.00000000045

Answers

1. $0.3 = 3 \times 10^{-1}$
2. $0.008 = 8 \times 10^{-3}$
3. $0.00105 = 1.05 \times 10^{-3}$
4. $0.0000357 = 3.57 \times 10^{-5}$
5. $0.0815 = 8.15 \times 10^{-2}$
6. $0.00001 = 1 \times 10^{-5}$
7. $0.004 = 4 \times 10^{-3}$
8. $0.324 = 3.24 \times 10^{-1}$
9. $0.000451 = 4.51 \times 10^{-4}$
10. $0.00000000045 = 4.5 \times 10^{-10}$

Test yourself 1.8

Convert the following numbers into standard form:

a) 10.5
b) 200
c) 308.6
d) 1230
e) One million
f) 0.3
g) 0.00045
h) 0.00000003
i) 0.005
j) 50.5

Converting numbers which look as though they are in standard form into standard form

Sometimes a number looks like it is in standard form but isn't. For example 12×10^{-4} looks like it is in standard form because it contains a power of 10. However the number 12 is too large.

This is how we convert numbers like this into standard form.

First convert the number 12 into a number between and including 1 and less than 10.

In this case 12 becomes 1.2. To keep this number the same as 12 we must multiply it by 10 like this $12 = 1.2 \times 10$

Hence, we can write $12 \times 10^{-4} = 1.2 \times 10 \times 10^{-4} = 1.2 \times 10^{-3}$

Pointer

$1.2 \times 10 \times 10^{-4}$ can be written as $1.2 \times 10^{1} \times 10^{-4}$. Now you add the powers of 10 so we obtain 1.2×10^{-3}.

Examples

1 125×10^{5} is the result for a chemistry calculation. Express this number in standard form.

Answer

1 $125 \times 10^{5} = 1.25 \times 10^{2} \times 10^{5} = 1.25 \times 10^{7}$

2 Express the number 250×10^{-3} in standard form.

Answer

2 $250 \times 10^{-3} = 2.5 \times 10^{2} \times 10^{-3} = 2.5 \times 10^{-1}$

Test yourself 1.9

The following numbers appear to be in standard form, but they aren't. Convert each number into standard form:

a) 15×10^{-4}
b) 18×10^{4}
c) 10×10^{3}
d) 125×10^{-3}
e) 0.1×10^{-3}
f) 0.001×10^{3}
g) 0.5×10^{4}
h) 0.125×10^{-3}
i) 18.5×10^{-2}
j) 0.002×10^{4}

Calculations involving numbers in standard form

Although you will be using a calculator to perform calculations using numbers in standard form, it is a useful check to be able perform simple calculations manually.

Here are some calculations that can be carried out without using a calculator.

Example

1 Work out the following calculation without using a calculator giving your final answer in standard form.

$3 \times 10^{5} \times 4 \times 10^{2}$

Answer

1 $3 \times 10^{5} \times 4 \times 10^{2} = 12 \times 10^{7}$

Notice that the answer is not in standard form.

The number 12 is changed to 1.2 and the power of 10 is increased by one to compensate. Note that in changing from 12 to 1.2 you are dividing by 10 and hence you need to multiply by 10 to compensate. Hence the need to increase the power of 10 by one.

Hence, $12 \times 10^{7} = 1.2 \times 10^{8}$

» Pointer

Multiply the ordinary numbers first (i.e. multiply 3 by 4) and then add the indices together to give the power to which the 10 is raised.

Always check that your final answer satisfies the criteria for a number being in standard form.

Example

2 Work out the following calculation without using a calculator giving your final answer in standard form.

$0.5 \times 10^{-5} \times 4 \times 10^{3}$

Answer

2 $0.5 \times 10^{-5} \times 4 \times 10^{3} = 2 \times 10^{-2}$

» Pointer

Note that this answer is in standard form and needs no adjustment.

Example

3 Work out the following calculation without using a calculator giving your final answer in standard form.

$4 \times 10^{-8} \times 4 \times 10^{-2}$

Answer

3 $4 \times 10^{-8} \times 4 \times 10^{-2} = 16 \times 10^{-10} = 1.6 \times 10^{-9}$

Pointer

16×10^{-10} needs adjusting as it is not in standard form. The value needs to be changed to 1.6 and the power of 10 needs to be increased by 1.

Example

4 Work out the following calculation without using a calculator giving your final answer in standard form:

$\frac{8 \times 10^{-3}}{2 \times 10^{-2}}$

Answer

4 $4 \times 10^{-3-(-2)} = 4 \times 10^{-1}$

Test yourself 1.10

Without using a calculator, work out the following calculations involving standard form, giving your final answer in standard form:

a) $3 \times 10^{6} \times 2 \times 10^{8}$

b) $4.5 \times 10^{2} \times 2 \times 10^{3}$

c) $4 \times 10^{-2} \times 4 \times 10^{5}$

d) $5.5 \times 10^{-3} \times 2 \times 10^{-3}$

e) $\frac{8 \times 10^{5}}{4 \times 10^{3}}$

f) $\frac{2 \times 10^{-3}}{1 \times 10^{4}}$

g) $\frac{3 \times 10^{-4}}{1.5 \times 10^{-2}}$

h) $\frac{5 \times 10^{-3}}{0.5}$

i) $\frac{4.5 \times 10^{-6}}{0.5 \times 10^{-3}}$

j) $\frac{1.5 \times 10^{4}}{15 \times 10^{-2}}$

1.7 Ratio and proportion

Ratio is a comparison between two or more numbers.

For example, the ratio of 3 moles to 2 moles means 3 moles compared to 2 moles. This can be expressed mathematically as 3 moles : 2 moles.

If the units are the same (i.e. moles in this case) then the units can be removed and the words compared with can be replaced with the ratio symbol ':'.

So this ratio can be written as 3:2.

Ratios can be cancelled by dividing by the largest number that divides into the numbers in the ratio.

So, a ratio 6:12 can be simplified to 1:2 by dividing both numbers by 6.

It is always important to simplify ratios.

Test yourself 1.11

Simplify each of the following ratios:

a) 2:6
b) 4:12
c) 6:4
d) 1.5:0.5
e) 10:30
f) 25:60
g) 10:1000
h) 25:100
i) 20:100
j) 5:25

In order to remove units from a ratio all the numbers forming the ratio must be in the same units. For example, if you had the ratio 200 g : 1 kg you would have to change the units so the ratio was expressed all in g or all in kg. Changing the units to grams, the ratio becomes 200 g : 1000 g and because the units are now the same, the units can be removed.

So we have 200:1000 which can be simplified to 1:5 by dividing by 200.

Test yourself 1.12

Express these ratios as simply as possible, without units:

a) 250 g : 2 kg
b) 1 mg : 10 g
c) 20 mg : 1 g
d) 25 g : 250 g
e) 400 mg : 20 g
f) 20 g : 1 kg
g) 4 mole : 8 mole
h) 0.01 mole : 1 mole
i) 0.2 mole : 0.01 mole
j) 0.8 mole : 0.2 mole

Using ratios to predict amounts in moles in a chemical reaction

The following chemical reaction shows how ammonia is manufactured from nitrogen and hydrogen:

$$N_2 + 3H_2 \longrightarrow 2NH_3$$

Although there is no number placed in front of N_2, the N_2 implies one molecule.

Hence using the equation we can see that one molecule of nitrogen reacts with 3 molecules of hydrogen to produce 2 molecules of ammonia.

Using moles rather than molecules, we can say that one mole of nitrogen reacts with 3 moles of hydrogen to produce 2 moles of ammonia.

We can also express the numbers of molecules or moles for products and reactants as a ratio like this:

1:3:2

Now, with ratios you can divide or multiply any of the numbers in the ratio by another number provided all the numbers in the ratio are treated the same way.

So, we can multiply all the numbers by 2 to give

$$2:6:4$$

or, we could divide the original ratio by 10 to give

$$0.1:0.3:0.2$$

Suppose we have the equation for the production of ammonia again

$$N_2 + 3H_2 \longrightarrow 2NH_3$$

If we want to find the number of moles of N_2 and H_2 to produce 0.3 moles of NH_3 we can use ratios in the following way:

$$N_2 + 3H_2 \longrightarrow 2NH_3$$

Ratio of $N_2 : H_2 : NH_3$ is 1:3:2

We need to change the last number in this ratio to 0.3 and alter the other numbers so that all the amounts are in the same ratio.

At the moment the last number is 2. We can change this to 1 by dividing the ratio by 2.

Hence, we have $1:3:2 = \frac{1}{2} : \frac{3}{2} : \frac{2}{2} = 0.5:1.5:1$

We can now change the last number from 1 to 0.3 by multiplying all the numbers in the ratio by 0.3 like this

$$0.5:1.5:1 = 0.5 \times 0.3 : 1.5 \times 0.3 : 1 \times 0.3 = 0.15:0.45:0.3$$

So the number of moles of N_2 and H_2 to produce 0.3 moles of NH_3 are 0.15 and 0.45 respectively.

Test yourself 1.13

a) Here is an equation for the reaction of magnesium with hydrochloric acid:

$$Mg(s) + 2HCl(aq) \longrightarrow MgCl_2(aq) + H_2(aq)$$

i. Find the number of moles of Mg needed to produce 4 moles of $MgCl_2$.

ii. Find the number of moles of HCl needed to produce 0.2 moles of H_2.

b) When aluminium burns in oxygen it forms aluminium oxide according this equation:

$$4Al(s) + 3O_2(g) \longrightarrow 2Al_2O_3(s)$$

i. Determine the number of moles of Al_2O_3 formed when 1 mole of Al is burnt completely in oxygen.

ii. It is desired to produce 0.5 moles of Al_2O_3. Determine the number of moles of oxygen needed for this reaction.

Direct proportion

Another way of comparing the way one quantity varies with another is by using direct proportion.

» Pointer
Note that there are no numbers in front of A, B, C or D. If there are no numbers, we take the number of moles of each of these to be one mole.

Take the following chemical reaction where the reactants A and B react together to give the products C and D according to the following equation:

$$A + B \longrightarrow C + D$$

From the equation, we can say that one mole of A reacts with one mole of B to give one mole of C and one mole of D.

Suppose we want to make the comparison between the amount of reactant A needed to produce product C we can see that the ratio in moles will be 1:1.

We can now write this under the equation like this:

$$\begin{array}{ccccccc} A & + & B & \longrightarrow & C & + & D \\ \text{1 mole} & & & & \text{1 mole} & & \end{array}$$

The quantities in this equation are in proportion. So if we double the number of moles of A the number of moles of C will double. If we halve the number of moles of A, the number of moles of C will halve. It is important to note that if we expressed the amounts of A and B in grams rather than in moles this would not happen as the masses are not in proportion. We therefore have to work in moles and then convert the moles back to grams if we want masses in grams.

Example

Butane burns completely in air according to the following equation:

$$2C_4H_{10} + 13O_2 \longrightarrow 8CO_2 + 10H_2O$$

If 2 tonnes of butane was burnt completely in air, what is the mass in tonnes of carbon dioxide produced? Note 1 tonne = 1000 kg.

Answer

As we are only concerned with the butane and carbon dioxide we can write the number of moles of each which react according to the equation like this:

$$\begin{array}{ccc} 2C_4H_{10} + 13O_2 & \longrightarrow & 8CO_2 + 10H_2O \\ \text{2 moles} & & \text{8 moles} \end{array}$$

We now need to find the M_r of butane and carbon dioxide by looking up the A_r of their constituent atoms in the periodic table.

M_r of butane = 58.0 g mol^{-1}

M_r of carbon dioxide = 44.0 g mol^{-1}

$$\begin{array}{ccc} 2C_4H_{10} + 13O_2 & \longrightarrow & 8CO_2 + 10H_2O \\ 2 \times 58.0\text{ g} & & 8 \times 44\text{ g} \\ = 116\text{ g} & & = 352\text{ g} \end{array}$$

» Pointer
You have now converted the amounts from moles to grams.

Now we find what mass of carbon dioxide 1 g of butane would produce.

To do this divide the masses by 116

So

$$\begin{array}{ccc} 2C_4H_{10} + 13O_2 & \longrightarrow & 8CO_2 + 10H_2O \\ \frac{116}{116} = 1\text{ g} & & \frac{352}{116} = 3.03\text{ g} \end{array}$$

Now as this is a ratio and both units are the same (i.e. grams) the units can be removed.

Hence we have $2C_4H_{10} + 13O_2 \longrightarrow 8CO_2 + 10H_2O$

1 3.03

Changing the units to tonnes, 1 tonne of butane would produce 3.03 tonnes of carbon dioxide.

Chapter 2 Handling data

2.1 Approximating numbers

Rather than give exact values of numbers, we often approximate them using a certain number of decimal places or significant figures and the following section shows how these approximations are made.

Decimal places

123.764 is given to 3 decimal places as there are three numbers given to the right of the decimal point.

Suppose we are asked to approximate the following number to three decimal places:

12.97547

Pointer
At the moment, this number is given to 5 decimal places. We need to change this so that only 3 numbers after the decimal point are shown.

To give a number to a specified number of decimal places you look at the figure in the next decimal place. In this case we would look at the fourth number to the right of the decimal point. If this number is less than 5, then the numbers up to and including the third place of decimals will stay the same. If the number is 5 or greater, the number for the third place of decimals will need to be increased by 1.

Examples

1 Write the number 4.56378 to the following numbers of decimal places

a) 4 d.p.

b) 3 d.p.

c) 2 d.p.

d) 1 d.p.

Answer

1 a) 4.5638

b) 4.564

c) 4.56

d) 4.6

2 Here is the result of a calculation in chemistry:

0.97598

Give this result to the following numbers of decimal places:

a) 1 d.p.

b) 2 d.p.

c) 3 d.p.

d) 4 d.p.

Answer

2 a) 1.0

b) 0.98

c) 0.976

d) 0.9760

Pointer

Notice the way that we write 1.0 rather than just 1. This is to show that we have written the number to one decimal place.

Test yourself 2.1

Write each of the following numbers to the number of decimal places specified in brackets:

(a) 10.565 (1)

(b) 123.9765 (3)

(c) 0.02557 (2)

(d) 3.9707 (3)

(e) 0.10455 (3)

(f) 0.00195 (3)

(g) 4.098 (2)

Significant figures

Many numbers in chemistry are written to a certain number of significant figures. In calculations you will often get an answer which has a full calculator display of numbers. The number of numbers a calculator can display depends on the make and model of the calculator. You are not expected to write all the numbers down as your answer. Instead you need to give your answer to a certain number of significant figures.

In calculations you will be given values which need to be put into an equation to calculate an unknown value. These values will be given correct to a certain number of significant figures.

Here are some values and the number of significant figures they are written to:

273 is written correct to 3 significant figures.

4200 is written correct to 2 significant figures.

8.31 is written correct to 3 significant figures.

0.0002 is written correct to 1 significant figure.

0.076 is written correct to 2 significant figures.

0.1005 is written correct to 4 significant figures.

Notice the following:

The counting for the figures starts at the first non-zero number.

You need to be able to approximate a number to a number with a certain number of significant figures. Here are some examples.

Example

Here is a number which appeared in the calculator display as a result of a calculation.

3.5609

Write the above number correct to the following numbers of significant figures:

a) One significant figure

b) Two significant figures

c) Three significant figures

d) Four significant figures

Answer

» Pointer
Note we usually abbreviate significant figures by s.f.

Remember to count the numbers starting with the first non-zero number. Once you reach the required significant figure (the first in this case) look at the number on the right to see if it is 5 or more. If it is then the number for the significant figure is increased by one. If it is less than 5 the number for the significant figure remains unaltered.

a) 3.5609 = 4 (correct to 1 s.f.)

b) 3.5609 = 3.6 (correct to 2 s.f.)

c) 3.5609 = 3.56 (correct to 3 s.f.)

d) 3.5609 = 3.561 (correct to 4 s.f.)

Example

Write each of the following numbers to 3 significant figures:

a) 1259

b) 14.919

c) 0.0003579

d) 0.009884

e) 1.8099×10^{-4}

Answer

a) The third significant figure in this number is 5 and since the number to the right of it is 9, the number 5 is increased to 6. In a large number like this the numbers after the required significant figure are replaced by a zero or zeroes.

Hence, 1259 = 1260 (correct to 3 s.f.)

Remember you don't start counting the numbers for significant figures, until the first non-zero number.

b) 14.919 = 14.9 (correct to 3 s.f.)

c) 0.0003579 = 0.000358 (correct to 3 s.f.)

d) 0.009884 = 0.00988 (correct to 3 s.f.)

e) $1.8099 \times 10^{-4} = 1.81 \times 10^{-4}$ (correct to 3 s.f.)

Test yourself 2.2

Express the following results of calculations correct to 3 significant figures:

a) 12.6666 c) 1.87888 e) 1.2058×10^8

b) 0.06356 d) 0.0008959 f) 2.03988×10^{-4}

2.2 Making estimates of the results of calculations

It is so easy to make a mistake when entering numbers into a calculator, so it is important to have an idea what the answer should be so that you can recognise a wrong answer. Making estimates is an important part of checking your answers.

If your examination involves multiple choice questions then you can sometimes make estimates of calculations and then choose the correct answer from your estimate.

Example

Estimate the value of: $2.8 \times 0.9 \times 3.4$, giving your answer to one significant figure.

Answer

First, start by writing all the numbers in the question to one significant figure. Note that with practice this can be done in your head.

Hence, $2.8 \times 0.9 \times 3.4 \approx 3 \times 1 \times 3 = 9$

Working this out accurately on a calculator $2.8 \times 0.9 \times 3.4 = 8.568 = 9$ (1 s.f.)

Pointer
The symbol ≈ means approximately equal to.

Example

Estimate the value of $\frac{2.68 \times 7.89 \times 12.00}{2.98 \times 6.07 \times 11.67}$, giving your answer correct to one significant figure.

Answer

$\frac{2.68 \times 7.89 \times 12.00}{2.98 \times 6.07 \times 11.67} \approx \frac{3 \times 8 \times 12}{3 \times 6 \times 12} = \frac{4}{3} = 1$ (1 s.f.)

Working this out accurately on a calculator $\frac{2.68 \times 7.89 \times 12.00}{2.98 \times 6.07 \times 11.67} = 1.2020$

$= 1$ (1 s.f.)

Pointer
Remember to cancel the numbers in the numerator (i.e. top part) with those in the denominator (i.e. bottom part) if possible.

Example

Estimate the value of $\sqrt{(2.5 \times 4.1 \times 2.9)}$, giving your answer correct to one significant figure.

Answer

$\sqrt{(2.5 \times 4.1 \times 2.9)} \approx \sqrt{(3 \times 4 \times 3)} = \sqrt{36} = 6$ (1 s.f.)

Working this out accurately on a calculator

$\sqrt{(2.5 \times 4.1 \times 2.9)} = 5.4521 = 5$ (1 s.f.)

Test yourself 2.3

Use estimation to find which of the answers A, B, or C is closest to the exact answer.

You should approximate each number in the calculation to one significant figure.

	Question	Answer A	Answer B	Answer C
1	3.45 × 2.78 × 0.09	0.8	1	0.08
2	12.56 × 1.87 × 0.45	6	26	13
3	120 ÷ 0.45	300	200	60
4	0.01 × 0.15 × 109	0.3	0.2	2
5	0.12 × 300 × 0.53	15	41	9
6	6.07 × 3.67 × 0.1	4	2	5
7	20.75 ÷ 6.98	0.3	4	3
8	0.01 × 145 × 35	40	400	450
9	6.5 × 0.3 × 0.01	0.01	0.02	0.2
10	65 ÷ 1050	0.07	0.1	0.001

2.3 Finding the arithmetic mean

The mean of a set of n numbers is the sum of the numbers divided by n.

For example, the mean of the set of numbers 12, 15, 21, 34 and 71 is

$$\frac{12 + 15 + 21 + 34 + 71}{5} = \frac{153}{5} = 30.6$$

The weighted mean

The element chlorine consists of two isotopes. One isotope has a mass of 35 whilst the other has a mass of 37. If a sample contained half the total number of atoms with a mass of 35 and half the number of atoms with a mass of 37 we could simply add these together and divide by 2 to give the mean mass which in this case would be $\frac{35 + 37}{2} = 36$.

However, it is found that one of the isotopes (i.e. the one with mass 35) is more abundant.

In fact in a sample there is 75.53% of chlorine-35 and 24.47% of chlorine-37.

This means the lighter atoms are much more common in a sample and this must be taken into account when calculating the mean. A mean which takes the abundance into account is called a weighted mean.

The weighted mean is calculated in the following way:

$$\text{Weighted mean} = \frac{75.53 \times 35 + 24.47 \times 37}{100} = 35.5 \text{ (3 s.f.)}$$

Example

What is the mean atomic mass for thallium, Tl if thallium consists of two isotopes, one having mass 205.1 and an abundance of 70.5% and the other with mass 203.1 and abundance 29.5%.

Answer

$$\text{Weighted mean} = \frac{70.5 \times 205.1 + 29.5 \times 203.1}{100} = 204.5 \text{ (4 s.f.)}$$

2.4 Units in chemistry

In chemistry most quantities we deal with have units and the main units we use are included in the following table. Chemists use a range of units, some based on the older CGS units (based on the centimetre, gram and second) and others based on SI units (based on the metre, kilogram, second and ampere). Here is a table showing the common units you will come across in chemistry.

Quantity	Unit	Symbol
Mass	gram or kilogram	g or kg
Amount of substance	mole	mol
Length	centimetre, metre	cm or m
Time	second	s
Electric current	ampere	A
Temperature	Celsius or kelvin	°C or K
Energy	joule	J
Electric charge	coulomb	C
Volume	cubic centimetre, cubic metre, cubic decimetre	cm^3, m^3, dm^3
Density	gram per cubic centimetre	$g\,cm^{-3}$
Pressure	pascal	Pa
Molar mass	grams per mole	$g\,mol^{-1}$

The units in the table above can have the following prefixes added in front of the unit.

For example, ms means 10^{-3} s.

Prefix	Symbol	Meaning
deci	d	10^{-1}
centi	c	10^{-2}
milli	m	10^{-3}
micro	µ	10^{-6}
nano	n	10^{-9}
kilo	k	10^{3}
mega	M	10^{6}
giga	G	10^{9}
tera	T	10^{12}

2.5 The ideal gas equation

The ideal gas equation is frequently used in chemistry. When dealing with this equation you must make sure that any values substituted into this equation are in the correct units. In this section we will look at this equation and see how to convert quantities that are not in the correct units.

The ideal gas equation is $pV = nRT$

where

p is the pressure of the gas in pascals (Pa)

V is the volume of the gas in m^3

n is the amount of gas in mol

R is a constant called the gas constant which has a value of 8.31 $J\,K^{-1}\,mol^{-1}$

T is the temperature in kelvin, K

Converting temperatures from Celsius °C to kelvin K

» Pointer

To convert from Celsius to kelvin you add 273 to the temperature in °C.

On the kelvin scale of temperature, 0 K = −273 °C. The kelvin scale is called the absolute temperature scale. Note that 0 K is the lowest temperature possible.

To change from °C to K we use the formula

$$K = °C + 273$$

Examples

Convert the following temperatures into kelvin:

a) 10 °C

b) 25 °C

c) 100 °C

d) −20 °C

Answers

a) 10 °C = (10 + 273)K = 283 K

b) 25 °C = (25 + 273)K = 298 K

c) 100 °C = (100 + 273) = 373 K

d) −20 °C = (−20 + 273) = 253 K

Converting volumes

In the ideal gas equation $pV = nRT$, the volume V must be in the units m^3. If you are given other units for the volume of the gas, they must be converted to m^3.

Converting from cm^3 to m^3

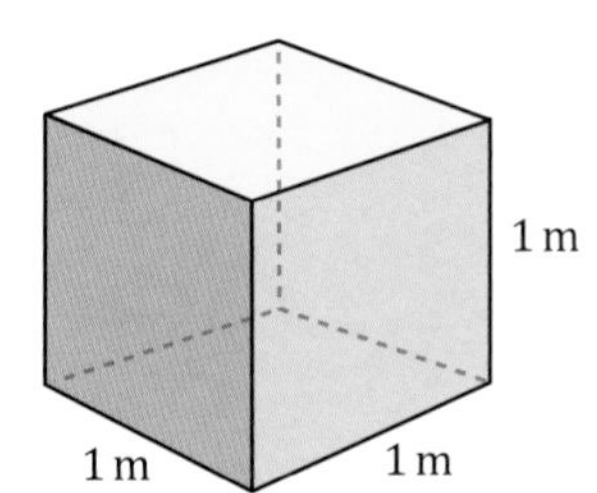

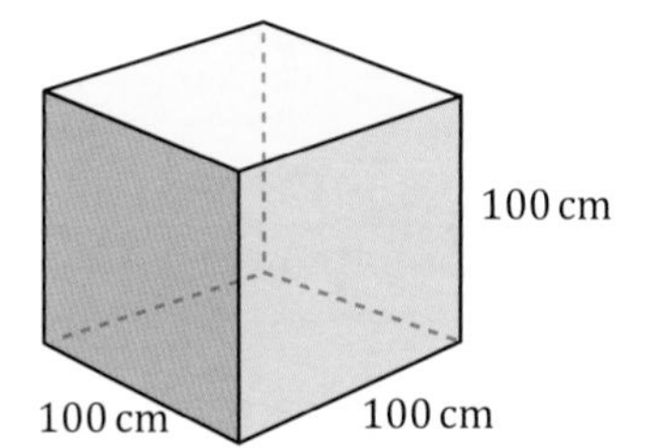

The above cubes have identical sides but they are measured in different units. The volumes of the two cubes are identical. The first cube has a volume of $1 \times 1 \times 1 = 1\ m^3$ and the second cube has a volume of $100 \times 100 \times 100 = 1\ 000\ 000\ cm^3$.

As the cubes have the same volume, we have

$$1\ m^3 = 1\ 000\ 000\ cm^3$$

Dividing both sides by 1 000 000 we obtain

$$\frac{1}{1\ 000\ 000}\ m^3 = 1\ cm^3$$

$$\text{Hence } 1\ cm^3 = \frac{1}{1 \times 10^6}\ m^3 = 1 \times 10^{-6}\ m^3$$

Example

Convert 250 cm^3 to m^3.

Answer

$250\ cm^3 = 250 \times 1 \times 10^{-6}\ m^3$

$= 2.5 \times 10^{-4}\ m^3$

Converting from dm^3 to m^3

1 dm means one decimetre (one tenth of a metre) (i.e. 0.1 m)

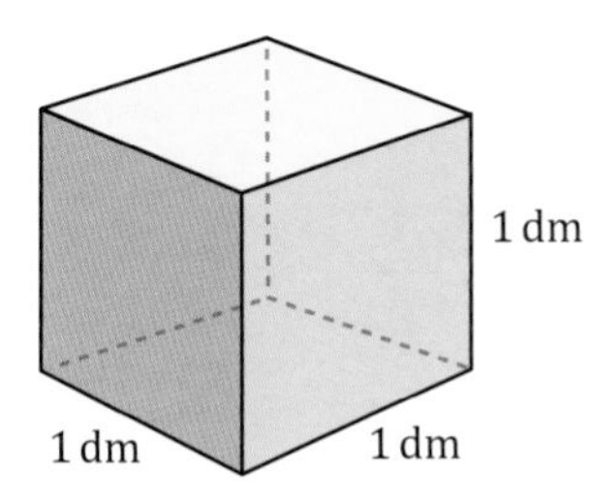

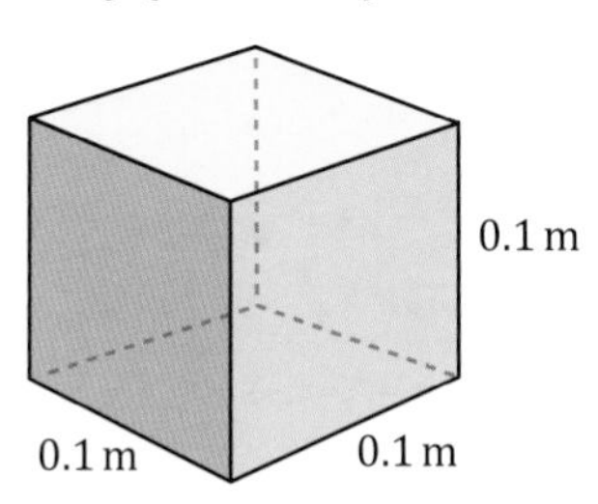

$1\ dm^3 = 0.1 \times 0.1 \times 0.1\ m^3$

$= 0.001\ m^3$

$= 1 \times 10^{-3}\ m^3$

So, to convert from dm^3 to m^3 you either multiply by 0.001 or by 1×10^{-3}

Example

Convert a volume of 5 dm^3 to m^3.

Answer

$5\ dm^3 = 5 \times 0.001\ m^3$

$= 0.005\ m^3$

Or

$5\ dm^3 = 5 \times 1 \times 10^{-3}\ m^3$

$= 5 \times 10^{-3}\ m^3$

Test yourself 2.4

1. Convert the following volumes from cm^3 to m^3 giving all the answers in standard form:
   (a) 5 cm^3
   (b) 250 cm^3
   (c) 0.5 cm^3
   (d) 1500 cm^3
2. Convert the following volumes from dm^3 to m^3.
   (a) 20 dm^3
   (b) 450 dm^3
   (c) 1800 dm^3
   (d) 0.1 dm^3

Chapter 3 Algebra

3.1 Symbols used in algebra

There are a number of symbols used in algebra and these are summarised in the following table. Some of these may be new to you.

Symbol	Meaning	Example
=	Equal	$N = 2$ (both values on either side of the equals sign are the same).
<	Less than	$n < 3$ (the value of n is smaller than 3). Note the value of n cannot be 3.
<<	Much less than	$m_e << m_p$ (the mass of an electron is much less than the mass of a proton).
>>	Much greater than	$m_p >> m_e$ (the mass of a proton is much greater than the mass of an electron).
>	Greater than	$n > 3$ (the value of n is greater than 3). Note the value of n cannot be 3.
∝	Proportional to	$y \propto x$ (the value of y is proportional to the value of x) if x doubles y doubles, if x halves y halves, etc.

3.2 Transposition of formulae

You will frequently have to rearrange a formula to change the subject. The subject of the formula is the quantity that needs to be found and it will appear on its own in the formula.

Take the following example:

$$n = \frac{m}{M}$$

» Pointer
Notice the use of upper case and lower case letters to mean different things in this formula. M is the molar mass or M_r and m is the mass of the substance.

As n is on its own on the left-hand side, it is the subject of the formula. Suppose we want to change the subject of the formula to m. We need to get rid of M on the right of the equation and we do this by multiplying both sides of the equation by M.

The equation now appears as follows:

$$Mn = \frac{Mm}{M}$$

» Pointer
We normally write the letters in an equation in alphabetical order. So here we would write Mn rather than nM.

We can now cancel the M in the top and bottom of the fraction on the right-hand side to give

$$Mn = m$$

This equation can be swapped around so that the left part appears on the right and vice versa to give

$$m = Mn$$

The equation has now been transposed so that m is the subject.

Pointer

Note that if the required subject of the equation is in the denominator, then multiply both sides by it.

If $n = \frac{m}{M}$ and we want to make M the subject we first multiply both sides by M to give

$$Mn = m$$

Now, we need to remove the n and this is done by dividing both sides by n.

So, $M = \frac{m}{n}$ and M is now the subject.

3.3 Representing a variable using more than one letter

In chemistry not all the variables in formulae use a single letter. For example, temperature is normally represented in a formula using the letter T. A change in temperature (i.e. a temperature difference) is represented by the variable ΔT. The Greek letter Δ (i.e. delta) is used to represent a change.

Sometimes the same letter needs to be used in a single formula to represent different things. For example, a quantity called entropy is always given the letter S in formulae. However, we need to attach other information to the 'S' to make it different.

Here is a formula you will come across later in the book in the topic on thermodynamics.

$$\Delta S_{system} = \Delta S_{products} - \Delta S_{reactants}$$

Suppose we want to make $\Delta S_{reactants}$ the subject of the equation shown above.

First add $\Delta S_{reactants}$ to both sides of the equation. This will has the effect of making $\Delta S_{reactants}$ positive and on the left-hand side of the equation like this

$$\Delta S_{system} + \Delta S_{reactants} = \Delta S_{products}$$

Now subtract ΔS_{system} from both sides and we have the required equation

$$\Delta S_{reactants} = \Delta S_{products} - \Delta S_{system}$$

Test yourself 3

1. The ideal gas equation is $pV = nRT$.
   Rearrange this equation so that each of the following is the subject of the equation:
   (a) V
   (b) n
   (c) T
   (d) p

2. The energy of electromagnetic radiation E is related to the frequency f by the following equation

$$E = hf$$

   where h is a constant called Planck's constant.

   Rearrange the above equation to make f the subject of the equation.

3. Here is the equation of a straight line

$$y = mx + c$$

   Rearrange the above equation so that m is the subject of the equation.

4. Make the bracketed symbol the subject of the equation; e.g. in question (a), write the equation in the form $\lambda =$.
   (a) $c = f\lambda$ (λ)
   (b) $c = \frac{n}{V}$ (V)
   (c) $Q = mc\Delta T$ (ΔT)
   (d) $n = \frac{V}{1000} \times c$ (V)
   (e) $n = \frac{V}{1000} \times c$ (c)
   (f) $E = hf$ (h)
   (g) $\Delta G = \Delta H - T\Delta S$ (ΔS)

3.4 Solving equations

Solving an equation involves finding the value of an unknown quantity. When this quantity is found and substituted back into the equation it makes the equation correct, meaning that the left-hand side of the equation has the same value as the right-hand side.

If the unknown quantity is the subject of the equation then all you need to do is to evaluate the other side of the equation by substituting the numbers into it.

Example

If $Q = mc\Delta T$, find Q if $m = 100$ g, $c = 4.18$ J g^{-1} K^{-1} and ΔT=6K.

Answer

$Q = mc\Delta T$

$Q = 100 \times 4.18 \times 6$

$Q = 2508$ J

» Pointer
Always write out the equation you are going to use before substituting the numbers.

» Pointer
In maths we do not always attach units to numbers. In chemistry, we nearly always have to consider units.

This equation finds heat energy, so Q is measured in Joules. You need to know the context of the question in order to know what unit to use.

If the unknown quantity is not the subject of the equation, then it is necessary to manipulate the equation. There are two ways of doing this:

1 Substitute the numbers into the equation and then manipulate.

2 Manipulate the equation to make the unknown quantity the subject and then substitute the numbers in.

The following example shows the two methods.

Example

The ideal gas equation is $pV = nRT$. Find the value of p if $n = 5$, $R = 8.31$, $V = 3$ and $T = 298$.

Answer

Method 1 Substituting the numbers into the equation and then manipulating.

$pV = nRT$

$p \times 3 = 5 \times 8.31 \times 298$

$p = \frac{5 \times 8.31 \times 298}{3}$

$= 4127.3$

$= 4130$ (3 s.f.)

Method 2 Manipulating the equation to make the unknown quantity the subject and then substituting the numbers in.

$pV = nRT$

$p = \frac{nRT}{V}$

$= \frac{5 \times 8.31 \times 298}{3}$

$= 4127.3$

$= 4130$ (3 s.f.)

» Pointer
As we only have to give the value of the pressure, we do not have to include any units.

3.5 Logarithms

What are logarithms?

Look at the following $10^3 = 1000$. This says that 10 to the power of 3 is equal to 1000.

In the above example the 10 is called the base and the 3 is called the logarithm of 1000 to the base 10. We write $3 = \log_{10} 1000$.

We have the following general rule:

If $X = b^p$ then by definition, $p = \log_b X$.

It is important to note the following:

$$\log_{10} 1 = 0 \text{ because } 10^0 = 1 \text{ and } \log_{10} 10 = 1 \text{ because } 10^1 = 10$$

Note also that you cannot find the log of zero or the log of a negative number.

Bases used for logarithms

There are two bases used for logarithms, base 10 and base e where e is the number 2.71828... In A-level Chemistry we only need to deal with logarithms having a base 10.

Log_{10} X is written simply as Log X, so the base is left out and it is assumed it is 10.

To find the log of a number using a calculator, simply press the button marked log and then enter the number and press =.

Using logs in relation to quantities that range over several orders of magnitude

You can see that logs are a bit complicated, so why are they used? Logarithmic scales are scales of measurement used where the value of a physical quantity uses intervals that correspond to orders of magnitude rather than a standard linear scale.

Suppose we wanted to draw a graph and instead of plotting a linear scale such as 0, 0.1, 0.2, 0.3, 0.4, 0.5 on one of the axes we needed to plot the following values 1, 10, 100, 1000, 10,000. If you try to draw this on a set of axes you will see the problem. As the scale varies in orders of magnitude (i.e. multiples of 10 in this case), it will be difficult as some of the values such as 0, 1, 10 and 100 are very close together when the scale has to reach 10,000.

To get around this problem we can use a logarithmic scale where we find the log of each of the numbers. We would now have a scale which covers log 1, log 10, log 100, log 1000 and log 10 000. Now, finding the logs of the numbers would give a scale covering 0, 1, 2, 3 and 4. We have now reduced the scale to a much more manageable size.

The presentation of data using a log scale is very useful when the data covers a large range of values. Later on in the book you will come across the pH scale, which is used to represent the acidity, neutrality or alkalinity of a solution. The quantity used to calculate pH (i.e. the hydrogen ion concentration) varies over many orders of magnitude, so this is why a logarithmic scale is used.

Chapter 4 Graphs

In chemistry and other subjects, we use graphs to display data graphically. If data is in a table it is hard to see patterns in it. When graphs are used, the patterns are much easier to see.

When we conduct experiments in chemistry, we often need to test the relationships between variables and we can do this by drawing graphs.

4.1 Straight line graphs

Straight line graphs are also called linear graphs and have an equation of the form

$$y = mx + c$$

Notice there is one y on the left-hand side of the equation.

m is the gradient (i.e. the steepness of the line) and c is the intercept on the y-axis.

It is important that in this equation there is only an x term, so there are no terms containing x^2, x^3, $\sqrt{x}$, $\frac{1}{x}$, etc.

The following equation is an equation of a straight line.

$$y = 2x - 3$$

Comparing this equation with

$$y = mx + c$$

You can see that the gradient, m is 2 and intercept on the y-axis, c, is −3.

Examples

1 Which of each of the following equations are equations of straight lines?

a) $y = 1.4x + 7$

b) $y = 4x^2 + 1$

c) $y = \frac{2}{x}$

d) $y = -4x$

e) $y = \sqrt{x}$

f) $2y = 8x + 6$

g) $5x + 2y = 3$

h) $y = \frac{4}{3}x - 2$

Answers

1 a) Yes: $y = 1.4x + 7$ is in the form $y = mx + c$

b) No: $y = 4x^2 + 1$ is not in the form $y = mx + c$ as it contains a term in x^2.

c) No: $y = \frac{2}{x}$ is not in the form $y = mx + c$

d) Yes: $y = -4x$ is in the form $y = mx + c$ with $c = 0$.

e) No: $y = \sqrt{x}$ is not in the form $y = mx + c$

f) Yes: $2y = 8x+6$ can be divided by 2 to give $y = 4x + 3$ which is in the form $y = mx + c$

g) Yes: $5x + 2y = 3$ can be rearranged to give
$2y = -5x + 3$ and this can be divided by 2 to give $y = -\frac{5}{2}x + \frac{3}{2}$ which is in the form $mx + c$

h) Yes: $y = \frac{4}{3}x - 2$ is in the form $y = mx + c$

Gradients of straight line graphs

Gradients have a sign and a value, so −0.6 is a gradient, as is 4.

The sign of the gradient

The gradient of a straight line can be positive (if y increases as x increases), negative (if y decreases as x increases) or zero (if the value of y stays the same as x increases).

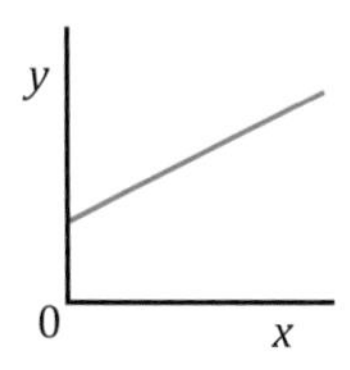

Fig 4.1 Positive gradient

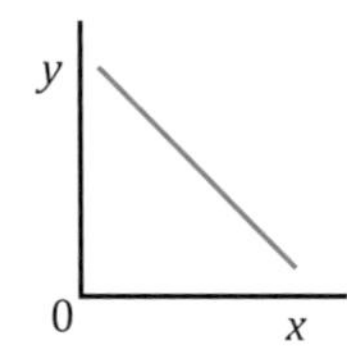

Fig 4.2 Negative gradient

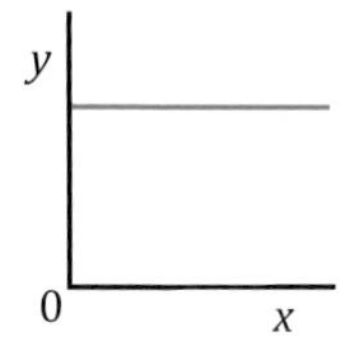

Fig 4.3 Zero gradient

Finding the gradient

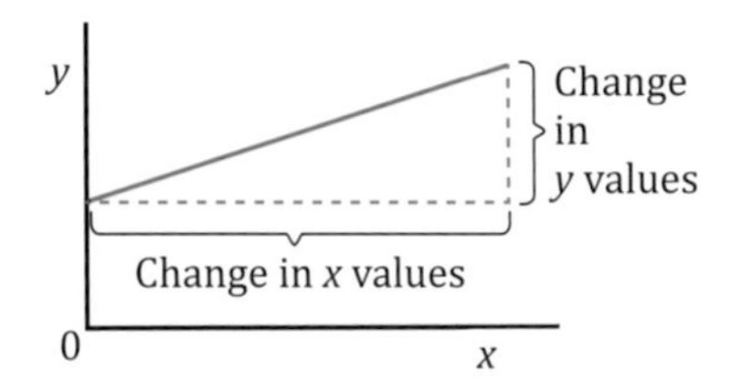

Fig 4.4

To find the size of the gradient, we draw a triangle like the one formed in Fig 4.4. We then find the distances shown on the diagram and use the following relationship for the gradient:

$$\text{Gradient, } m = \frac{\text{change in the } y\text{-values}}{\text{change in the } x\text{-values}}$$

Note that we have to look at which way the line slopes to determine whether the gradient of the line is positive or negative. We only include the sign, if the gradient is negative. If the gradient is positive, say 4, then we do not write +4.

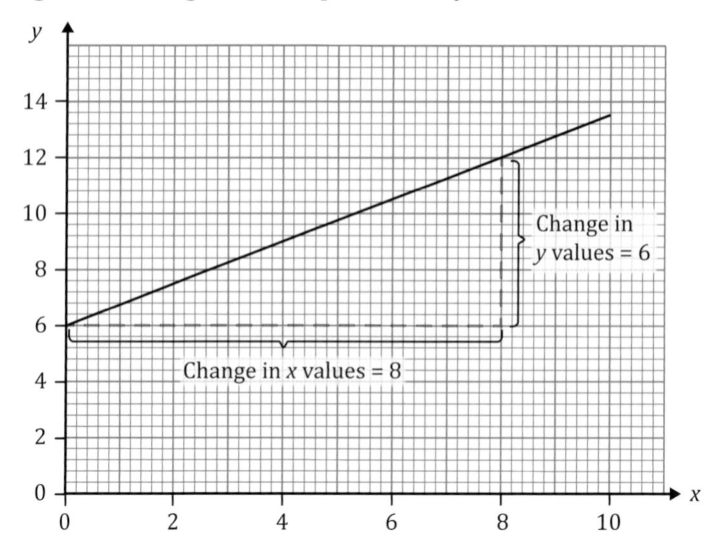

Fig 4.5

Fig 4.5 shows a straight line graph. In order to find the gradient of this graph we need to find a convenient place to construct a triangle. We try to find a place where the graph cuts through one of the corners of one of the large squares on the grid. It is more accurate if the triangle is made as large as convenient. You can see that the points (0, 6) and (8, 12) are convenient points to start and finish the triangle. From the graph you can see that the two distances are 6 and 8.

Notice that for the line y increases as x increases so the gradient is positive.

$$\text{We then use, gradient, } m = \frac{\text{change in the } y\text{-values}}{\text{change in the } x\text{-values}} = \frac{6}{8} = \frac{3}{4} \text{ or } 0.75$$

4.2 Linear relationships

A linear relationship is one in the form $y = mx + c$, where x and y are the variables and m is the gradient of the line and c is the intercept of the line on the y-axis.

If a graph of a quantity plotted on the y-axis against a quantity plotted on the x-axis is a straight line, then the two quantities are linearly related.

Take the following straight line, for example

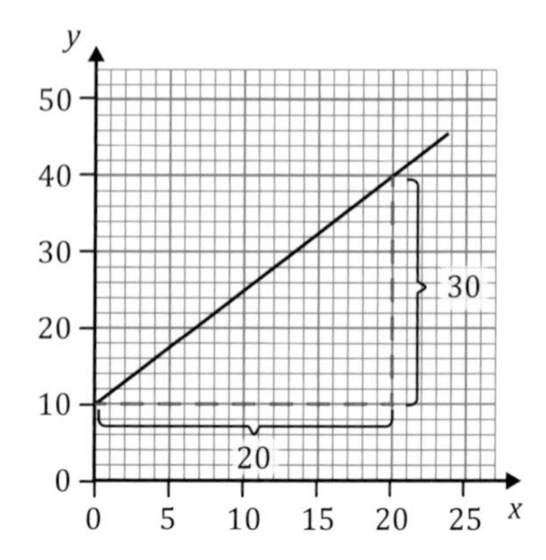

The gradient, m, is found forming a triangle.

$$\text{Gradient} = \frac{\text{change in the } y\text{-values}}{\text{change in the } x\text{-values}} = \frac{30}{20} = 1.5$$

Intercept on the y-axis, $c = 10$

Hence equation of the straight line

$$y = mx + c$$

$$y = 1.5x + 10$$

Pointer

The values of m and c obtained from the graph are substituted into the general equation for a straight line.

Straight line graphs showing direct proportion

If the straight line passes through the origin (0,0) or simply O, then the quantity on the y-axis is directly proportional to the quantity on the x-axis. Directly proportional means that if, for example, the quantity on the x-axis doubles then the quantity on the y-axis doubles. In other words, whatever you multiply the quantity on the x-axis by, the quantity on the y-axis will go up by the same factor.

Drawing a straight line graph from experimental data

Most graphs in chemistry arise from experimental data (i.e. each data point is the result of a measurement of combination of measurements). When the points are plotted they often involve a degree of scatter owing to some uncertainty in the measurements. If the points look as though they suggest a straight line then we need to draw a straight line of best fit.

Line of best fit

To draw a line of best fit:

- Use a transparent ruler to draw a line that matches the gradient suggested by the data points.
- Ensure that the points are scattered equally above and below the line.

Calculating the rate of change from a graph showing a linear relationship

Graphs used to determine the rate of change of a quantity plotted on the y-axis always have time along the x-axis. Such graphs can be used to determine the rate of change of the quantity on the y-axis.

In the following example a graph of volume of oxygen in m^3 is plotted on the y-axis and the time in seconds is plotted on the x-axis.

The rate of change of the quantity (i.e. volume of oxygen) on the y-axis is determined from the gradient of the graph.

Example

Hydrogen peroxide (H_2O_2) decomposes to form oxygen (O_2) and water (H_2O) according to the equation

$$2H_2O_2 \longrightarrow 2H_2O + O_2$$

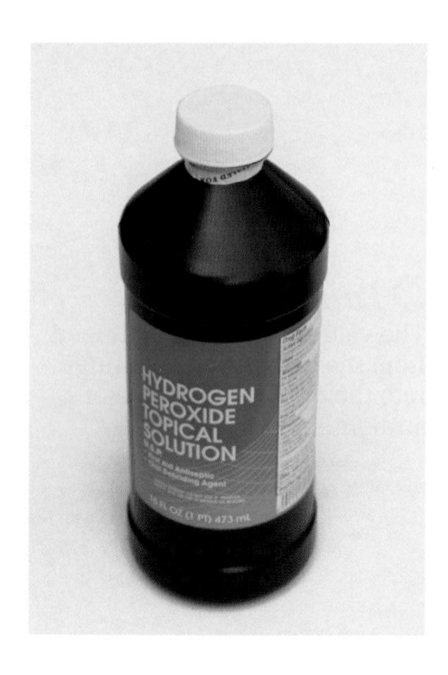

An experiment to find the rate of decomposition of H_2O_2 gave the following results

Time/s	0	50	100	150	200	250	300
Volume O_2/cm^3	0	5.0	10.0	14.8	19.0	22.5	25.0

Plot these results and determine the initial rate of production of oxygen from your plot.

Answer

The points are plotted. Make sure that you label both axes and include the units.

Notice the correct way that this is done. First include the name of the quantity and then use a / symbol and write the units after this. So on the y-axis we would use

Volume/cm^3

When you plot the points you will notice that not all the points lie on a straight line.

They are in an almost straight line initially but start to bend.

Hence, we need to draw a straight line up to about 200 s and then a smooth curve after that. As the start of the graph is a straight line we can say that the gradient and hence the rate of change of O_2 is constant.

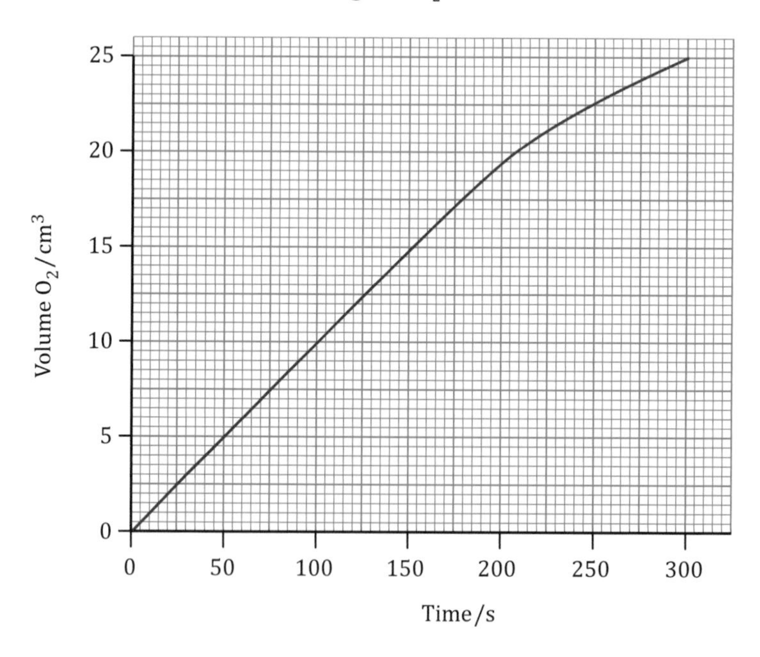

To find the gradient, we need to draw a triangle in the straight section of the graph.

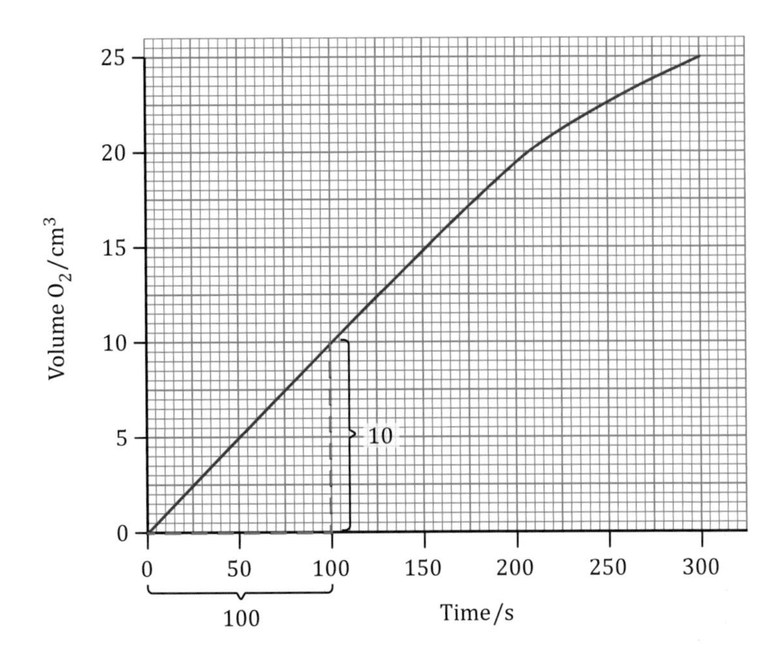

Hence

Initial rate of production of oxygen = gradient = $\frac{10}{100}$ = 0.1 cm^3 s^{-1}

» Pointer

Note that we could have used a larger triangle but it is more convenient to use a triangle where the line cuts the corner of one of the grid squares.

4.3 Drawing and using the slope of a tangent to a curve to determine rate of change

Many quantities in chemistry do not vary linearly with each other so their graphs are not straight lines. Instead their graphs are curves so the gradient is no longer constant and it depends on where on the graph it is measured. Look at the graph below and you can see that the gradient at the point x_1 can be found by drawing a tangent (i.e. a straight line which skims the curve) and then finding the gradient of the line. You can see that the gradient at point x_2 is larger than at x_1.

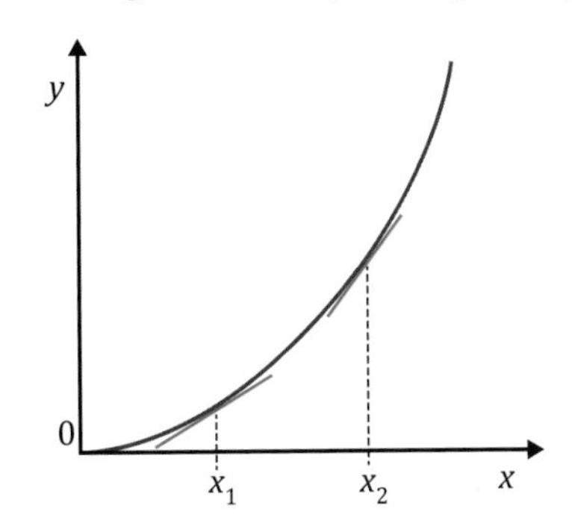

Test yourself 4

1. Find the gradient of the following straight line graphs (remember to include the sign if the gradient is negative). Give your answers as decimals correct to 2 decimal places.

a)

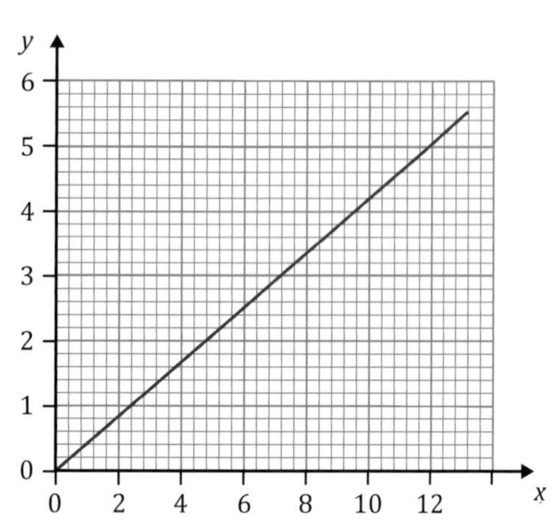

b)

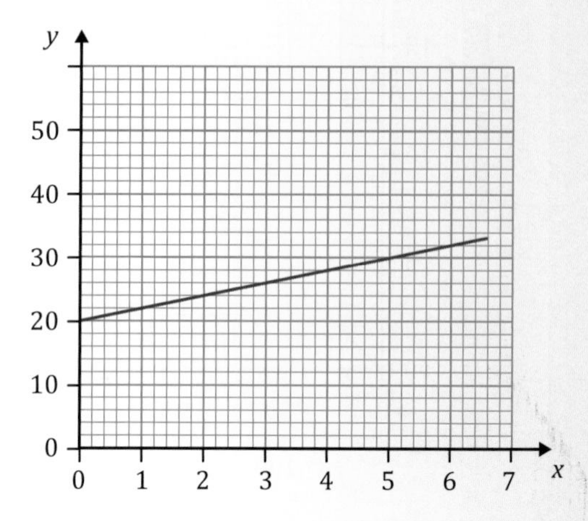

c)

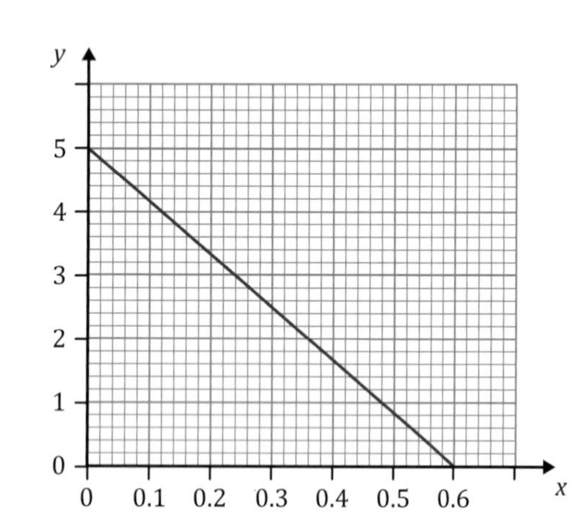

d)

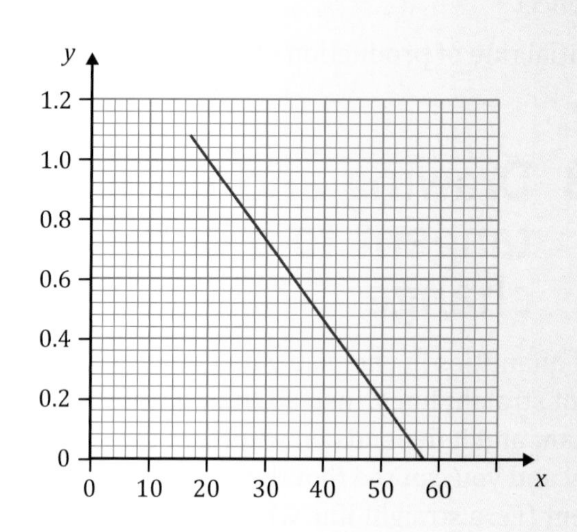

e)

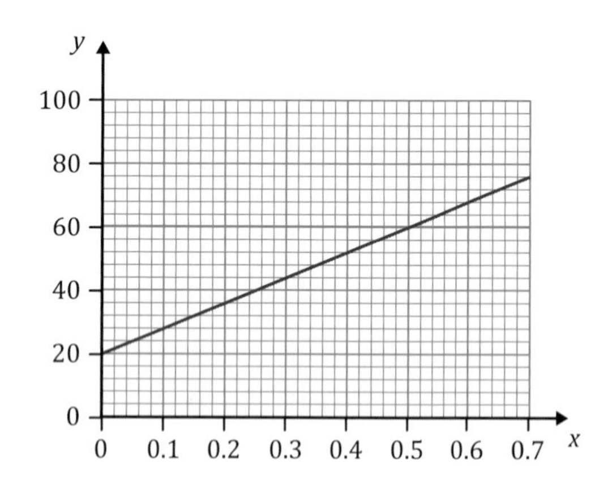

f)

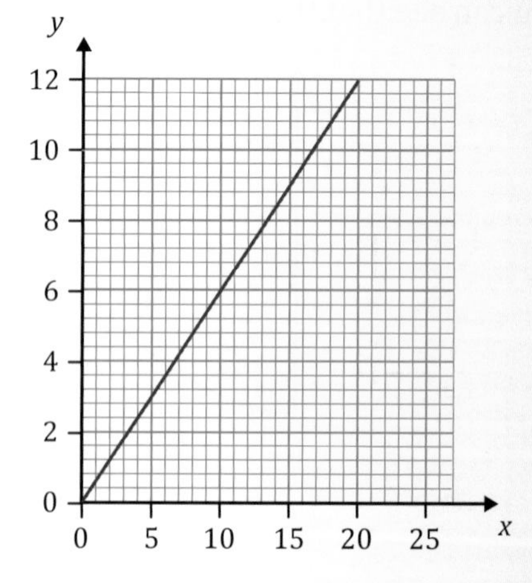

❷ Find the equation of the following straight lines giving your answer in the form

$y = mx + c$

a)

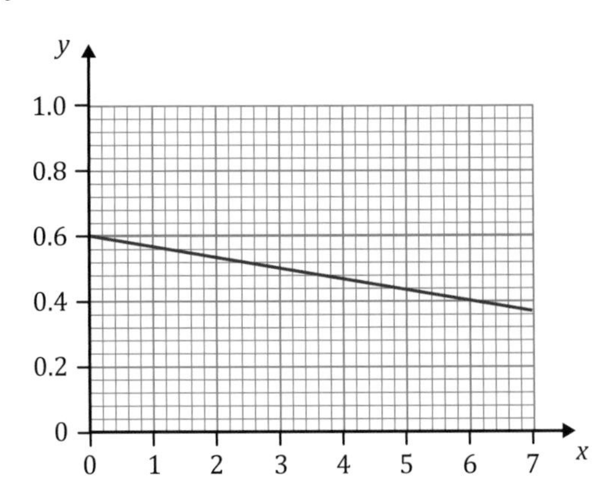

b)

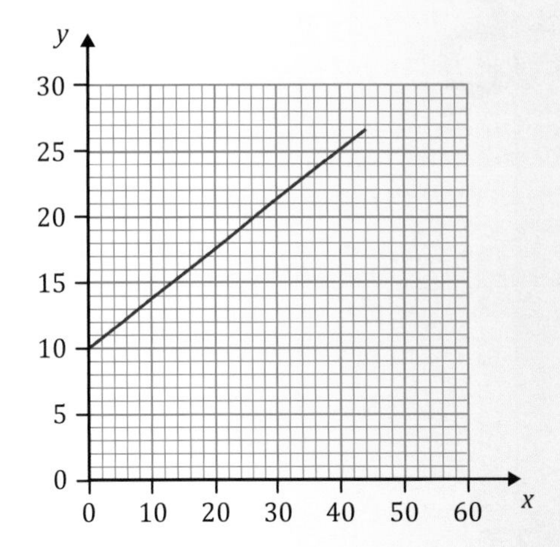

c)

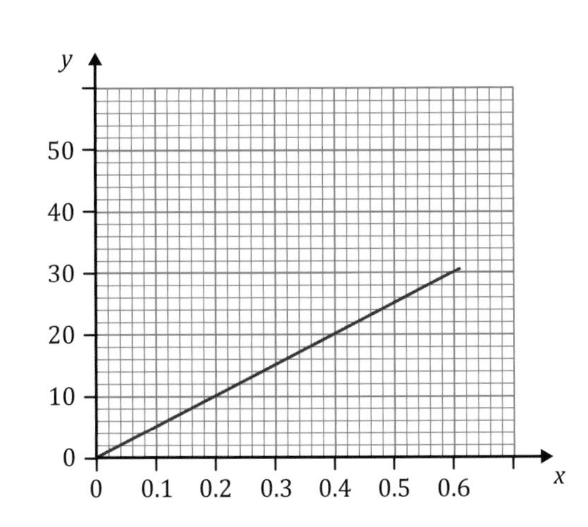

d)

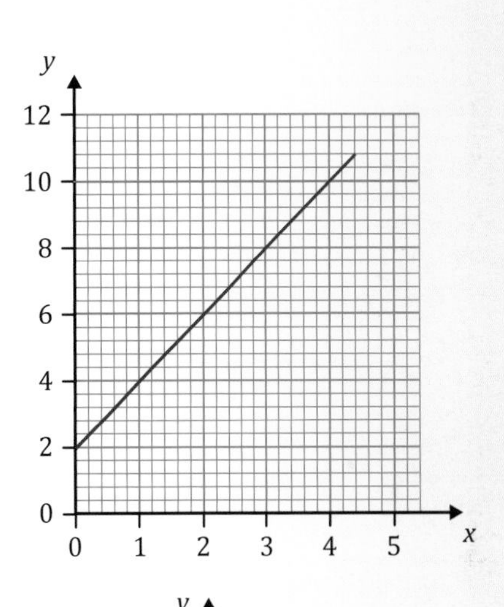

e)

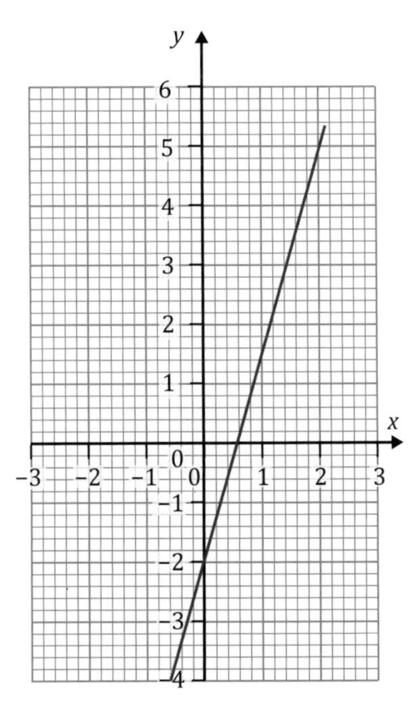

f)

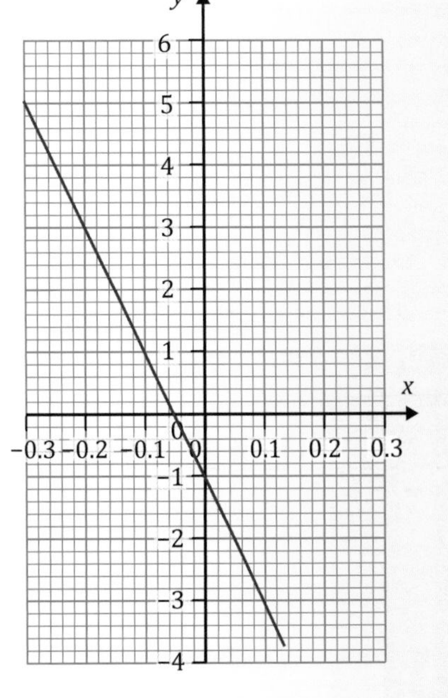

Chapter 5 Geometry

You need a knowledge of regular 2D and 3D shapes as well as an appreciation of angles for A-level Chemistry because you need to be able to picture the structure of atoms or molecules and understand the angles involved in structures.

5.1 2D shapes

2D shapes are flat and lie in the plane of the paper (i.e. they can be drawn accurately on a flat piece of paper).

Square (4 equal sides)

Rectangle
(2 pairs of equal sides)

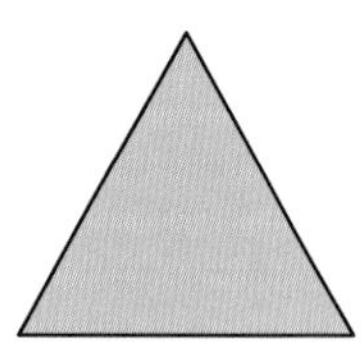

Equilateral triangle
(all sides are equal and all angles are 60°)

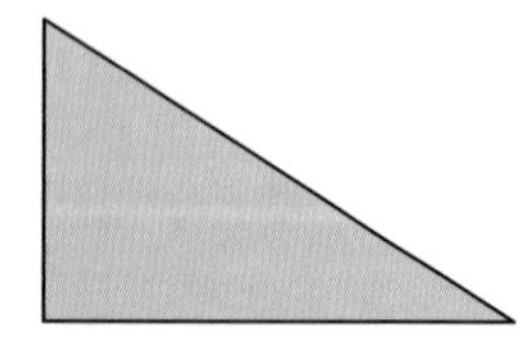

Right-angled triangle
(contains one angle of 90°)

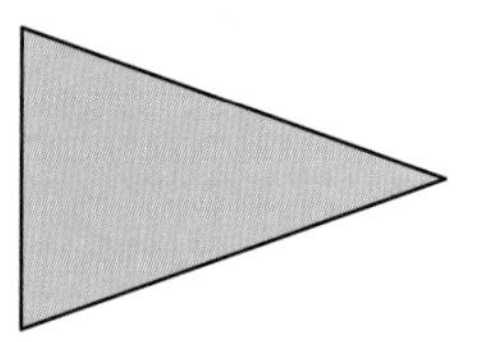

Isosceles triangle
(two equal sides and base angles are equal)

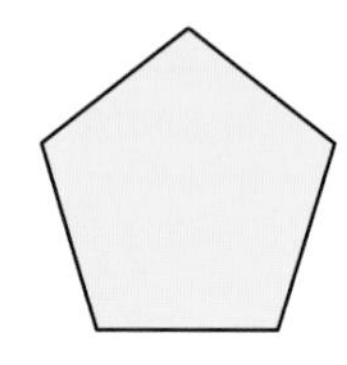

Regular pentagon
(5 equal sides and angles)

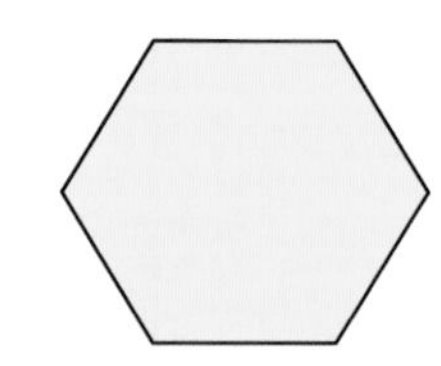

Regular hexagon
(6 equal sides and angles)

5.2 3D shapes

3D shapes cannot be represented accurately on a flat surface. Many molecules have 3D structures and it is important to be able to imagine how they look in 3D and to be able to draw a diagram to show the structure.

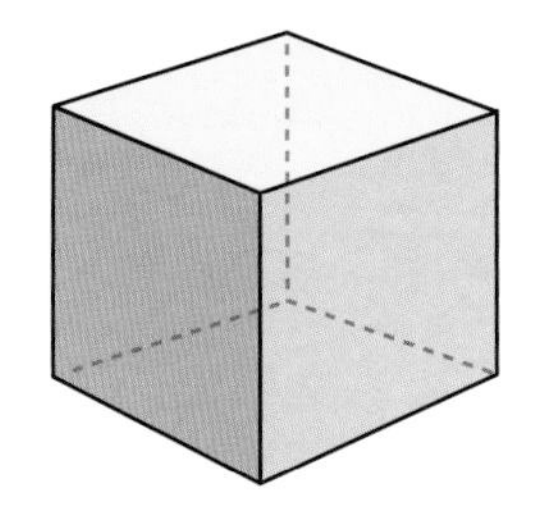

Cube
(8 vertices and 6 faces)

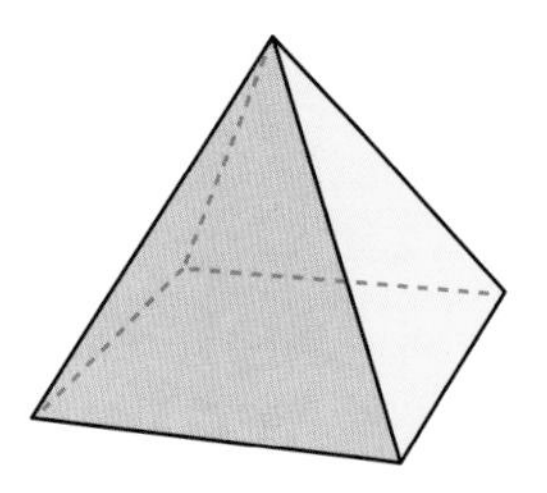

Pyramid
(square-based pyramid)
(5 vertices and 5 faces)

>> **Pointer**
Note that the vertices are the corners and the faces are the flat sides.

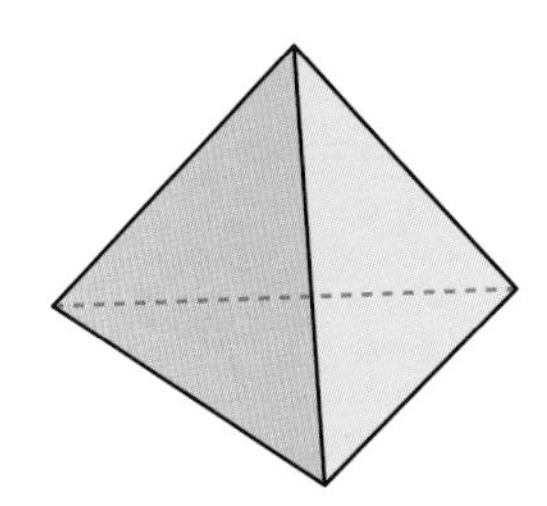

Tetrahedron
(triangular-based pyramid)
(4 vertices and 4 faces)

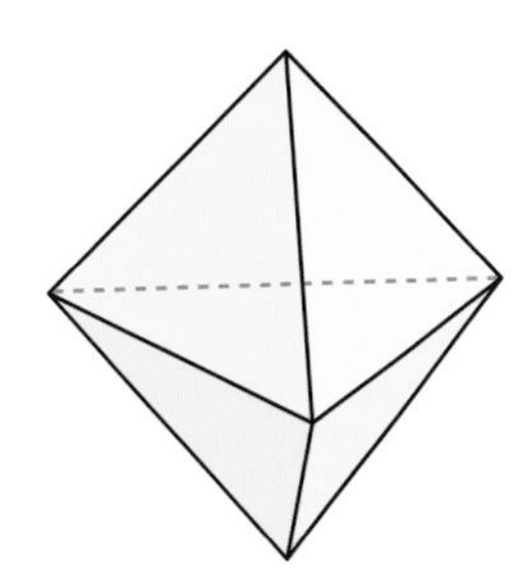

Trigonal bi-pyramid
(5 vertices and 6 faces)

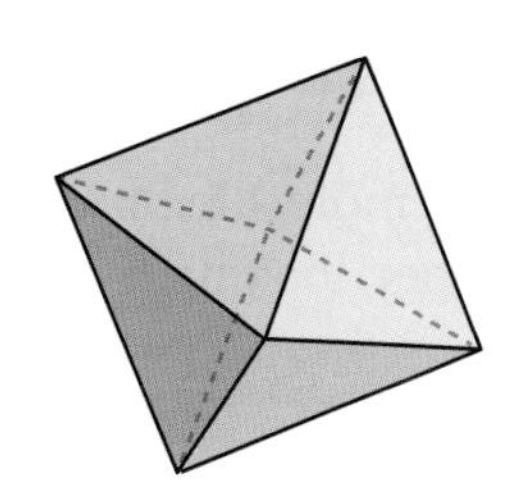

Octahedron
(6 vertices and 8 faces)

5.3 Angles

There are a number of angle rules which will help you understand about bond angles and they are described here.

Angles on a straight line

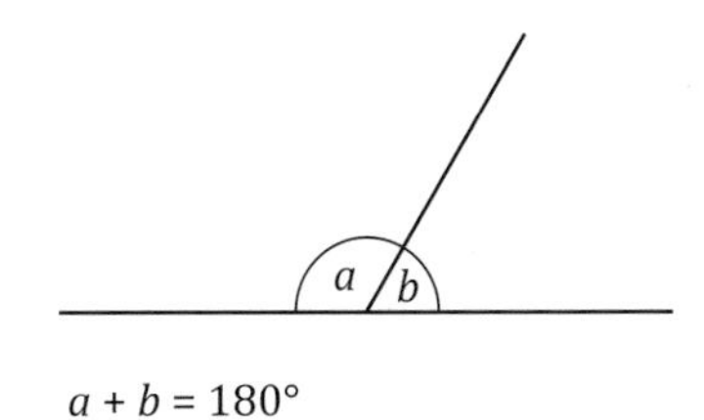

$a + b = 180°$

The angles on a straight line add up to 180°.

Angles at a point

The angles at a point add up to 360°.

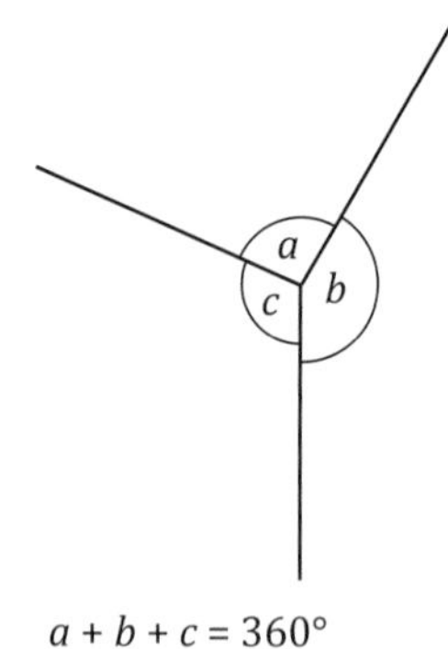

$a + b + c = 360°$

If three lines (which could represent bonds) act at a point and they need to be equally spaced, then the angles will be $\frac{360}{3} = 120°$.

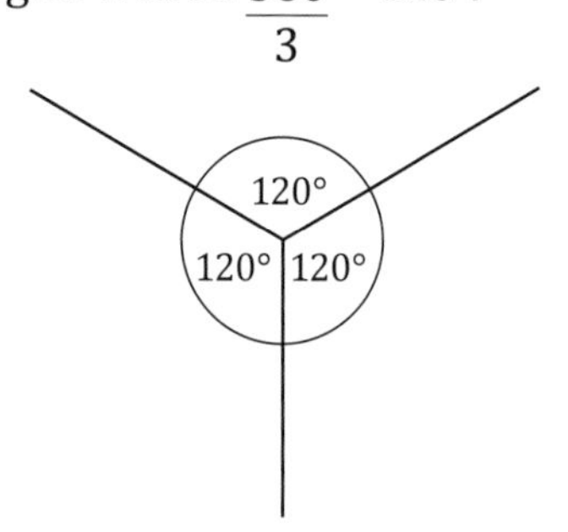

5.4 Viewing molecules in 2D and 3D

» Pointer
All the bonds and atoms in this molecule are in the same plane.

The following molecule boron trichloride is a trigonal planar molecule and when drawn on paper, looking from above the molecule is the following shape:

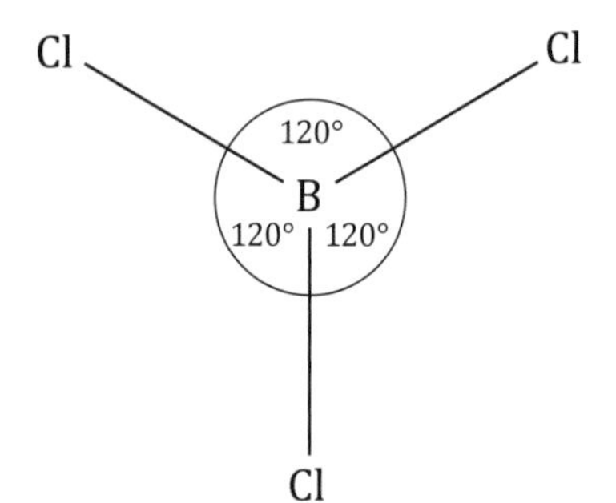

» Pointer
This view is like looking at the molecule drawn on a piece of paper positioned at right-angles to the page you are now looking at.

Another way we can draw this molecule is to put the B and one of the Cl atoms in the plane of the paper. One of the other Cl atoms would then be at an angle going into the paper and the other Cl atom would be at an angle coming out of the paper. This can be drawn in the following way:

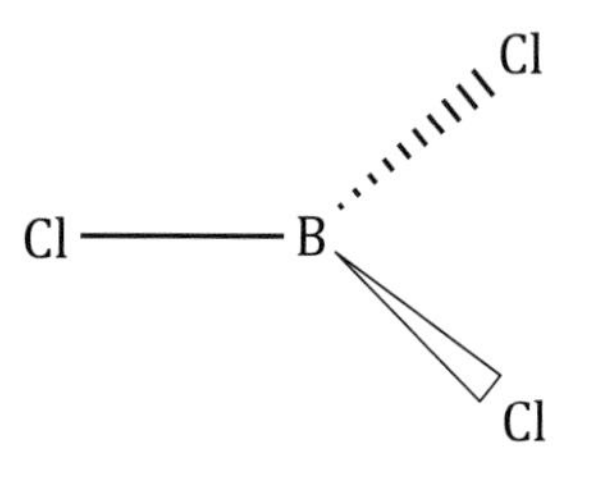

Note we can tell the direction of a bond using the following guide:

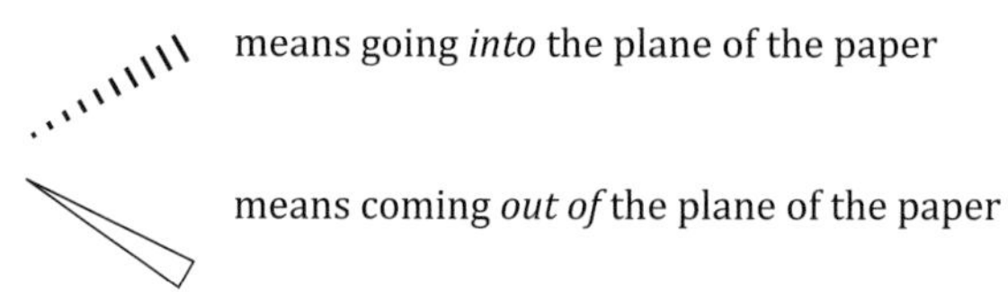

Tetrahedral-shaped molecules

A tetrahedron is another name for a triangular-based pyramid.

The methane molecule CH_4 has a tetrahedral structure with each of the hydrogen atoms situated at the vertices and the single carbon atom situated in a central position.

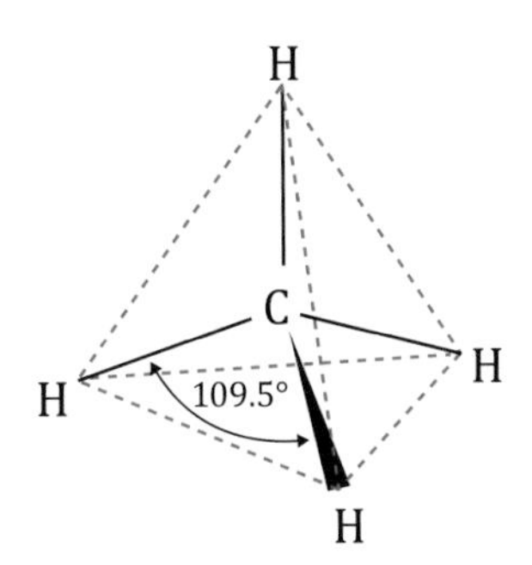

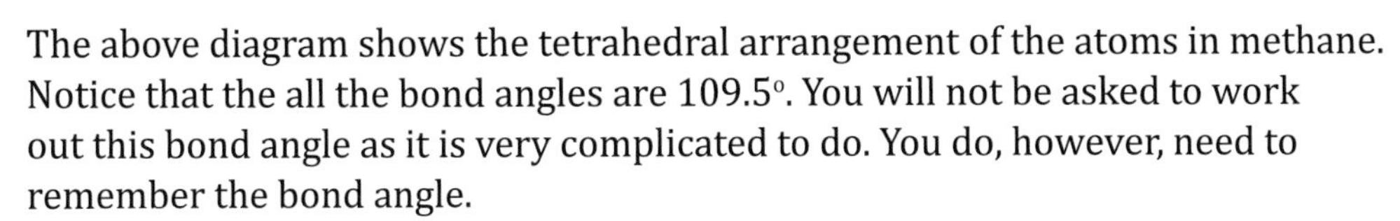

The above diagram shows the tetrahedral arrangement of the atoms in methane. Notice that the all the bond angles are 109.5°. You will not be asked to work out this bond angle as it is very complicated to do. You do, however, need to remember the bond angle.

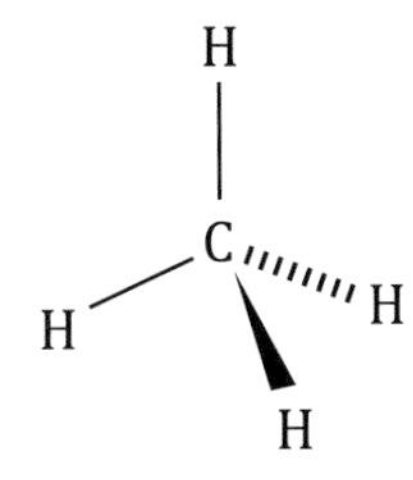

This shows the orientation of the bonds in CH_4.

Pointer

Remember that the wedge-shaped bond is used to show it comes out of the paper and the dotted bond shows it goes into the paper.

Octahedral-shaped molecules

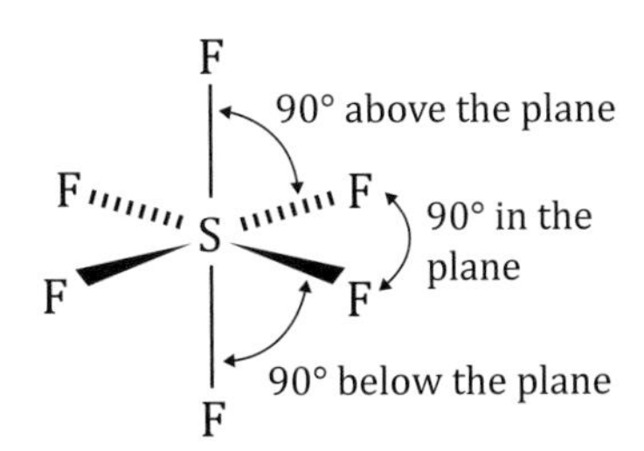

Sulfur hexafluoride (SF_6) is an octahedron-shaped molecule

The shapes of molecules and why they have these shapes will be looked at in a later chapter.

Test yourself 5

1. A molecule is shown in the diagram below. The lines represent bonds and the capital letters represent atoms. Determine the bond angle marked a in the diagram if all angles are equal.

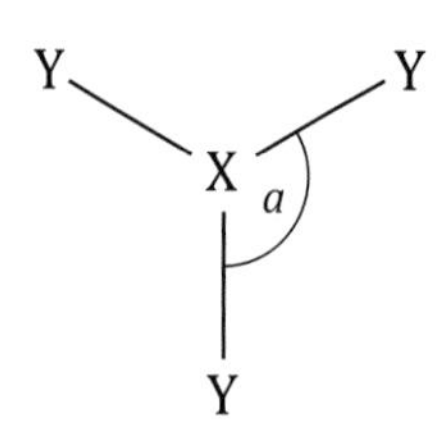

2. Draw a diagram for each of the following three-dimensional shapes:
   a) Cube
   b) Tetrahedron
   c) Pyramid
   d) Trigonal bi-pyramid
   e) Octahedron

3. The molecule sulfur hexafluoride SF_6 has the 4 fluorine atoms in the same plane and 2 fluorine atoms at right-angles to the plane. The molecule shape is shown in the following diagram:

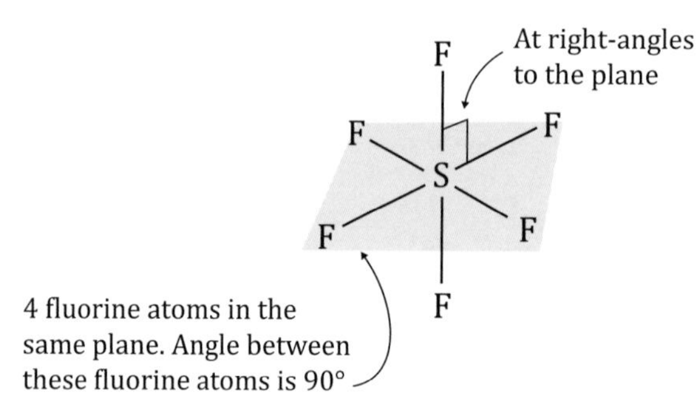

   Which one of the following describes the shape of this molecule?
   A Bi-pyramidal
   B Trigonal planar
   C Tetrahedral
   D Octahedral

4. The diagram below shows the shape of a XeF_4 molecule.

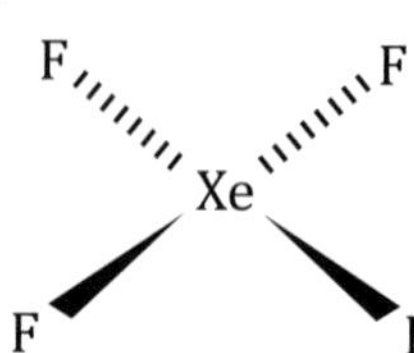

   If the angle between the bonds is 90°, which one of the following describes the shape of the molecule?
   A Tetrahedral
   B Square planar
   C Linear
   D Octahedral

Chapter 6 The amount of a substance

6.1 Relative atomic mass, A_r

Atoms have extremely small masses (typically from 10^{-24} to 10^{-22} grams), so instead of measuring the masses of atoms in grams, we compare the mass of an atom to the mass of a ${}^{12}_{6}C$ atom (i.e. an isotope of carbon atom having 6 protons and 6 neutrons in its nucleus). This gives us the relative atomic mass, A_r.

Relative atomic mass, A_r is defined as follows

$$\text{Relative atomic mass, } A_r = \frac{\text{Average mass of one atom of an element}}{\frac{1}{12}^{\text{th}} \text{ of the mass of one atom of carbon-12}}$$

Note: relative atomic mass has no units.

Pointer

The average mass of the atom is used because there may be several isotopes which will have different masses.

6.2 Definition of a mole

In chemistry we measure the amount of a substance in moles.

A mole of a substance is the amount of the substance that contains the same number of particles as there are carbon atoms in 12 g of carbon-12.

The mass of one mole of an element is its relative atomic mass in grams. The relative atomic mass of an element can be obtained from the periodic table. You will notice that for many elements this value will not be an integer (i.e. a whole number). This is because most elements consist of several isotopes with different masses, so the actual relative mass is a weighted average (i.e. mean) of all the isotopes.

Pointer

Always use the number of decimal places for values of A_r in the periodic table given in your specification/exam. Print out a copy from the exam board website and keep using it.

Finding the mass of one mole of an element

Look up the relative atomic mass in the periodic table.

This will be the mass in grams of one mole of the element.

Examples

One mole of carbon has a mass of 12.0 grams.

One mole of chlorine has a mass of 35.5 grams.

One mole of sulfur has a mass of 32.1 grams.

Pointer

Note that the relative atomic mass (A_r) for each element is looked up in the periodic table.

Finding the masses of more than one mole of an element

You simply multiply the A_r by the number of moles.

Examples

Find the masses in grams of:

a) 2 moles of copper

b) 0.5 moles of calcium

c) 4 moles of carbon

d) 1.5 moles of neon

e) 0.7 moles of sodium

Answers

a) 2 moles of copper = 2 × 63.5 g = 127 g

b) 0.5 moles of calcium = 0.5 × 40.1 = 20.05 g

c) 4 moles of carbon = 4 × 12.0 = 48.0 g

d) 1.5 moles of neon = 1.5 × 20.2 = 30.3 g

e) 0.7 moles of sodium = 0.7 × 23.0 = 16.1 g

» *Pointer*

Definitions for A_r and M_r need to be remembered as they are often asked for in exam questions.

6.3 Relative molecular mass M_r

Relative molecular mass, M_r is defined as follows

$$\text{Relative molecular mass, } M_r = \frac{\text{Average mass of one molecule of an element}}{\frac{1}{12}^{\text{th}} \text{ of the mass of one atom of carbon-12}}$$

Note: relative molecular mass has no units.

The relative molecular mass of a compound is the mass of one mole of the compound.

Take carbon dioxide (CO_2) for example. One mole of CO_2 consists of one mole of carbon atoms and two moles of oxygen atoms.

Hence the M_r for CO_2 will be (1 × 12.0) + (2 × 16.0) = 44.0

» *Pointer*

The multiplication must be done before the addition. The brackets have been added here to emphasise this.

Examples

Molecule	Symbol	Calculation of M_r using the A_r of component atoms	M_r
Hydrogen	H_2	2 × 1.0	2.0
Methane	CH_4	(1 × 12.0) + (4 × 1.0)	16.0
Water	H_2O	(2 × 1.0) + (1 × 16.0)	18.0
Ammonia	NH_3	(1 × 14.0) + (3 × 1.0)	17.0
Ethanol	C_2H_5OH	(2 × 12.0) + (6 × 1.0) + (1 × 16.0)	46.0

6.4 Relative formula mass

Ionic compounds do not consist of molecules as such but instead consist of ions. In the case of ionic compounds we use the term relative formula mass rather than relative molecular mass.

» *Pointer*

Relative molecular mass and relative formula mass both have the symbol M_r and no units.

The calculation for relative formula mass is the same as for working out the relative molecular mass.

Example

Find the relative formula mass of sodium chloride (NaCl).

Answer

One mole of sodium chloride contains one mole of sodium ions and one mole of chloride ions.

Note we can take the relative atomic mass of the ions to be the same as the relative atomic mass of the atoms.

Hence relative formula mass of sodium chloride = 23.0 + 35.5 = 58.5.

Examples

Compound	Symbol	Calculation of M_r using the A_r of component atoms	M_r
Sodium hydroxide	NaOH	23.0 + 16.0 + 1.0	40.0
Hydrochloric acid	HCl	1.0 + 35.5	36.5
Calcium carbonate	$CaCO_3$	40.1 + 12.0 + (3 × 16.0)	100.1
Sodium carbonate	Na_2CO_3	(2 × 23.0) + 12.0 × (3 × 16.0)	106.0
Silver nitrate	$AgNO_3$	108.0 + 14.0 + (3 × 16.0)	170.0

6.5 Molar mass (*M*)

Molar mass has the symbol *M* and is the mass of one mole of a substance (element or compound) in the units of g mol^{-1}.

For example, carbon has a molar mass of 12.0 g mol^{-1}.

Carbon dioxide has a molar mass of 12.0 + 2 × 16.0 = 44 g mol^{-1}.

» Pointer

The molar mass *M* and the relative molecular mass M_r are the same numerically. M_r has no units and *M* has units of g mol^{-1}.

6.6 Formula linking mass, molar mass and number of moles

The following formula linking mass, molar mass and number of moles is extremely important and will be used throughout the AS/A2 course and must be remembered.

$$\text{Number of moles, } n = \frac{m}{M}$$

where *m* is the mass of the substance in g and *M* is the molar mass of the substance in g mol^{-1}.

You must be able to rearrange this formula to make *M* or *m* the subject of the formula.

Maths help

Transpose the equation $n = \frac{m}{M}$

a) to make m the subject of the formula,

b) to make M the subject of the formula.

Answer

a) $n = \frac{m}{M}$

Multiplying both sides of the equation by M (note that doing this will remove the M in the denominator on the right-hand side).

$Mn = m$

Note that the sides can be swapped around to give

$m = Mn$

b) $n = \frac{m}{M}$

Multiplying both sides by M gives

$Mn = m$

Dividing both sides by n gives

$M = \frac{m}{n}$

Pointer

Note that as M is to be the subject of the equation it needs to be moved to the top of the equation by multiplying both sides of the equation by M.

Examples

1 State which one of the following contains the greatest number of moles:

A 32 g of oxygen atoms

B 27 g of water molecules

C 3 g of hydrogen atoms

D 88 g of carbon dioxide molecules

Answer

1 A $n = \frac{m}{M} = \frac{32}{16} = 2$ moles

B $n = \frac{m}{M} = \frac{27}{18} = 1.5$ moles

C $n = \frac{m}{M} = \frac{3}{1} = 3$ moles

D $n = \frac{m}{M} = \frac{88}{44} = 2$ moles

Answer = C

2 State which one of the following contains the least mass:

A 1 mole of carbon dioxide, CO_2

B 1.5 moles of methane, CH_4

C 2 moles of water, H_2O

D 3 moles of ammonia, NH_3

Answer

We rearrange $n = \frac{m}{M}$ so that m is the subject of the equation

Rearranging gives $m = nM$

A $m = 1 \times 44 = 44$ g

B $m = 1.5 \times 16 = 24$ g

C $m = 2 \times 18 = 36$ g

D $m = 3 \times 17 = 51$ g

Hence answer = B

» *Pointer*

To change the subject of the equation $n = m/M$ to make m, the subject of the equation, multiply both sides of the equation by M. The M on the right-hand side will cancel with the M in the denominator.

» *Pointer*

M, the molar mass, is determined for each of the molecules. Use the periodic table to obtain the figures needed.

6.7 Finding the percentage by mass of an element in a compound

To find the percentage by mass of an element in a compound, first find the mass of one mole of the compound. Then find the number of moles of the element in the compound and multiply it by the mass of one mole of the element. Find the fraction by dividing these two numbers, making sure that the molar mass is on the bottom of the fraction. The fraction is then multiplied by 100, which changes the fraction into a percentage.

$$\text{\% by mass of an element in a compound} = \frac{\text{mass of the element}}{\text{mass of one mole of the compound}} \times 100$$

Note: the mass of the element can be found by multiplying the A_r by the number of atoms in the compound.

Example

1 Show that the percentage by mass of sodium in sodium sulfate, Na_2SO_4 is 32.4%.

Answer

1 M_r of $Na_2SO_4 = (2 \times 23.0) + (1 \times 32.1) + (4 \times 16.0)$

$= 142.1$

There are two atoms of Na in one molecule of Na_2SO_4.

2 atoms of sodium have a mass of $2 \times 23.0 = 46.0$

$$\text{Percentage by mass of sodium} = \frac{\text{mass of sodium}}{\text{mass of sodium sulfate}} \times 100$$

$$= \frac{46}{142.1} \times 100$$

$$= 32.4\%$$

The Avogadro constant (L or N_A)

One mole of any substance contains the same number of particles. So, one mole of carbon atoms (i.e. 12.0 g) would contain the same number of particles as one mole of ammonia molecules (17.0 g).

» Pointer

Some examination boards use N_A rather than L to represent the Avogadro constant. Also both 6.022 and 6.02 are used. Please check with your exam board for the appropriate figure.

One mole of any substance contains a certain number of particles. This number is called the Avogadro constant, L or N_A, and is $6.022 \times 10^{23}\ \text{mol}^{-1}$.

It is important to note that particles can mean atoms, molecules, electrons or ions.

For example, a mole of copper atoms (i.e. 63.5 g) would contain the same number of particles as a mole of electrons. Both would contain $6.022 \times 10^{23}\ \text{mol}^{-1}$ particles.

6.8 Finding the number of particles in *n* moles of a substance

To find the number of particles in n moles of a substance you multiply the number of moles n, by the Avogadro constant (L or N_A).

$$\text{Number of particles in } n \text{ moles} = n \times L \quad \text{or} \quad n \times N_A$$

Examples

1 Calculate the number of particles in:

a) 2 moles of oxygen atoms

b) 1.5 moles of carbon dioxide molecules

c) 0.5 moles of calcium ions

Answer

1 a) Number of particles in 2 moles of oxygen atoms

$= 2 \times 6.022 \times 10^{23} = 1.2044 \times 10^{24}$

b) Number of particles in 1.5 moles of carbon dioxide molecules

$= 1.5 \times 6.022 \times 10^{23} = 9.033 \times 10^{23}$

d) Number of particles in 0.5 moles of calcium ions

$= 0.5 \times 6.022 \times 10^{23} = 3.011 \times 10^{23}$

2 Calculate the mass of ammonia, NH_3, which contains the same number of molecules as there are molecules in 54 g of water, H_2O.

» Pointer

The molar mass of water is the sum of the relative atomic masses of all the atoms that make up an H_2O molecule.

Answer

2 Molar mass of $H_2O = (2 \times 1.0) + (1 \times 16.0)$

$= 18.0\ \text{g mol}^{-1}$

Number of moles in 54 g of $H_2O = \frac{54}{18} = 3$

Pointer

This means that one mole of ammonia has a mass of 17 g.

Molar mass of $NH_3 = (1 \times 14.0) + (3 \times 1.0)$

$= 17.0\ \text{g mol}^{-1}$

The same number of moles of any substance, contains the same number of particles.

This means the number of moles of NH_3 must equal the number of moles of H_2O if they are to contain equal numbers of particles.

Hence the number of moles of NH_3 = 3.

As the molar mass of NH_3 = 17.0 g mol^{-1}, 3 moles of NH_3 have a mass of 3 × 17.0 = 51.0 g.

3 Write down the letter corresponding to the number of moles of each element in 88 g of carbon dioxide, CO_2, which has an M_r of 44.

	C	O
A	2	2
B	1	2
C	2	4
D	1	0.5

Answer

3 Number of moles of CO_2, $n = \frac{m}{M} = \frac{88}{44} = 2$

2 moles of CO_2 can be written as $2CO_2$.

Hence, there will be 2 moles of C and 4 moles of O.

Answer = C

Pointer

Note that the molar mass M can be replaced in this equation by M_r as they are the same numerically.

Test yourself 6

1. Find the relative molecular mass for each of the following molecules:
   (a) Carbon monoxide (CO)
   (b) Phosphorus pentachloride (PCl_5)
   (c) Propane (C_3H_8)
   (d) Ethanoic acid (CH_3COOH)
   (e) Ethanol (C_2H_5OH)
2. Find the relative formula mass of each of the following ionic compounds:
   (a) Sodium sulfate (Na_2SO_4)
   (b) Potassium hydroxide (KOH)
   (c) Hydrochloric acid (HCl)
   (d) Sodium hydroxide (NaOH)
   (e) Potassium iodide (KI)
   (f) Sodium carbonate (Na_2CO_3)
3. If the Avogadro constant is 6.02×10^{23} mol^{-1}. Find the number of atoms in 2 mol of carbon dioxide.
4. Butane (C_4H_{10}) burns completely in air to form carbon dioxide and water.
   (a) Write a balanced equation for the complete combustion of butane.
   (b) Find the mass of 5 mol of butane.

5. Solid sulfur is a yellow crystalline solid as shown below.

Sulfur exists as a lattice of S_8 molecules consisting of a ring of eight sulfur atoms.

A sample of sulfur contains 0.0150 mol of S_8 molecules. How many atoms of sulfur would there be in this sample?

6. Anhydrous sodium thiosulfate has the formula $Na_2S_2O_3$. Find the percentage by mass of sulfur in this compound.

7. It has been found that the human body contains around 0.025 g of iodine molecules, I_2.

If the Avogadro constant is $6.022 \times 10^{23}\ mol^{-1}$, calculate the number of iodine atoms in 0.025 g of I_2.

8. The relative atomic mass of tin is 118.7.

(a) Give a definition of the term relative atomic mass.

(b) A sample of pure tin has a mass of 234 g.

(i) Giving your answer to three significant figures, determine the number of moles of tin in this sample.

(ii) Giving your answer to three significant figures, determine the number of tin atoms in this sample.

Chapter 7 Balancing and using chemical equations to predict masses

7.1 Writing and balancing equations

You will have experienced writing and balancing chemical equations as part of your GCSE work. It is very important to be able to do this for AS/A2 Chemistry.

Example

Write a balanced equation for the burning of methane in a plentiful supply of air.

Answer

You need to know that when methane is burnt in a plentiful supply of air, the products are carbon dioxide and water.

Firstly write down the unbalanced equation showing the reactants and products:

$$CH_4 + O_2 \longrightarrow CO_2 + H_2O$$

Notice that one of the reactants (i.e. oxygen) appears twice in the products. When this happens it is best to balance this element last.

So balancing the carbon you can see that there is one atom on both sides, so it is balanced for carbon.

Looking at the hydrogen, there are 4 atoms in the reactants and only 2 atoms in the products so a 2 is put in front of the H_2O to balance the hydrogen:

$$CH_4 + O_2 \longrightarrow CO_2 + 2H_2O$$

Finally, we balance the oxygen. There are two atoms in the reactants and four in the products. If we put 2 in front of the oxygen in the reactants this will balance the oxygen.

Hence we have:

$$CH_4 + 2O_2 \longrightarrow CO_2 + 2H_2O$$

7.2 Stoichiometry

The stoichiometry of an equation tells us the ratio of amounts of reactants and products in a chemical reaction.

Take the following reaction as an example:

$$CH_4 + 2O_2 \longrightarrow CO_2 + 2H_2O.$$

The stoichiometry for the above reaction tells us that there is a ratio of 1:2:1:2 of methane, oxygen, carbon dioxide and water molecules.

Once an equation has been balanced correctly, the stoichiometry of the equation can be used to work out the molar amounts or amounts in grams for the reactants and products.

Calculations involving moles of substances

Suppose we want to determine the number of moles of ammonia produced when 1.5 moles of nitrogen (N_2) react with excess hydrogen (H_2) to produce ammonia (NH_3).

We first need to write a balanced equation in order to determine the stoichiometry:

$$N_2 + 3H_2 \longrightarrow 2NH_3$$

The stoichiometry tells us that there is a 1:3:2 ratio of molecules of nitrogen, hydrogen and ammonia. We can write the number of moles under the equation like this

Pointer
The number of molecules can be replaced by the number of moles.

N_2	+	$3H_2$	$\longrightarrow$	$2NH_3$
1 mole		3 moles		2 moles

Now we can multiply or divide these quantities by any number as long as we do it to all the reactants and products.

As we are asked to find the amounts for 1.5 moles of nitrogen, we multiply the moles of reactants and products by 1.5.

N_2	+	$3H_2$	$\longrightarrow$	$2NH_3$
1 × 1.5		3 × 1.5	$\longrightarrow$	2 × 1.5
= 1.5 mol		= 4.5 mol		= 3 mol

Hence we can see that 3 moles of ammonia would be produced.

Calculations involving masses of substances

The equation $n = \frac{m}{M}$ can be used to turn mole amounts of substances into masses in grams as the following examples show.

Examples

1 Magnesium nitrate undergoes thermal decomposition on heating to form magnesium oxide, nitrogen dioxide and oxygen according to the following equation:

$2Mg(NO_3)_2(s) \longrightarrow 2MgO(s) + 4NO_2(g) + O_2(g)$

A sample of magnesium nitrate was thermally decomposed to give 0.654 g of magnesium oxide.

a) Calculate the number of moles of MgO in 0.654 g of magnesium oxide.

b) Calculate the number of moles of NO_2 gas produced from the sample of magnesium nitrate.

c) Calculate the total number moles of gas produced from the thermal decomposition of this sample.

Answer

1 a) Using $n = \frac{m}{M}$

$$n = \frac{0.654}{24.3 + 16.0} = \frac{0.654}{40.3} = 0.0162 \text{ moles}$$

b) We know that 2MgO = 0.0162 moles

From the equation, $4NO_2$ is produced, which is double the amount of magnesium oxide.

Hence 2 × 0.0162 = 0.0324 moles of NO_2 is produced.

c) According to the equation 4 moles of $NO_2(g)$ and 1 mole of $O_2(g)$ of gaseous products would be produced.

$2Mg(NO_3)_2(s)$	$\longrightarrow$	$2MgO(s)$	+	$4NO_2(g)$	+	$O_2(g)$
2 moles		2 moles		4 moles		1 mole

However, 0.0162 moles of MgO are produced so we have:

$2Mg(NO_3)_2(s)$	$\longrightarrow$	$2MgO(s)$	+	$4NO_2(g)$	+	$O_2(g)$
0.0162 moles		0.0162 moles		0.0324 moles		0.0081 moles

Hence number of moles of gaseous products

= 0.0324 + 0.0081

= 0.0405 moles

2 Chloroethane, C_2H_5Cl, is produced by reacting ethene (C_2H_4) with hydrogen chloride (HCl).

	C_2H_4	+	HCl	$\longrightarrow$	C_2H_5Cl
M_r values	28.0		36.5		64.5

Calculate the maximum possible (theoretical) mass of chloroethane that could be obtained from 50 g of ethene.

Answer

2 The stoichiometry of the equation tells us that there is a 1:1:1 ratio of molecules of ethene, hydrogen chloride and chloroethane. The ratio of moles is the same (i.e. 1:1:1).

The equation tells us that 28.0 g of ethene reacts with 36.5 g of hydrogen chloride to give 64.5 g of chloroethane.

Now we can divide the amounts by any number provided that all the amounts (i.e. reactants and products) are divided by the same number.

» Pointer

As *M* and *m* are given to 3 s.f. the answer is given to 3 s.f. Answers should normally be given to 3 s.f. unless you are told otherwise in the question.

» Pointer

We know that 0.016 moles of MgO are produced, so we write this in under the equation and then keeping all the amounts in their correct ratio; write the other amounts in moles for the reactants and products.

» Pointer

Look carefully at the stoichiometry of the equation. You can see that $Mg(NO_3)_2$, MgO, NO_2 and O_2 are in the ratio of 2:2:4:1. If we let the number of moles of $Mg(NO_3)_2$ be 0.0162, then we need to double this to find the number of moles of NO_2 and halve it to find the number of moles of O_2.

» Pointer

Use the state symbols (i.e. the (g) in $O_2(g)$) to determine which of the products are gaseous.

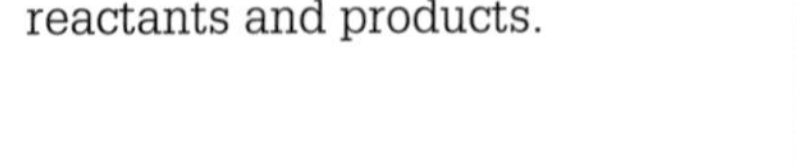

Pointer
You can multiply or divide the relative molecular masses or molar masses by any number provided you do this to all the reactants and products.

It would be useful to know the amounts for 1 g of ethene and this can be found by dividing all the amounts by 28.0 like this:

$$C_2H_4 + HCl \longrightarrow C_2H_5Cl$$

$$\frac{28.0}{28.0} = 1\text{ g} \qquad \frac{36.5}{28.0} = 1.3\text{ g} \qquad \frac{64.5}{28.0} = 2.3\text{ g}$$

To find the amounts for 50 g of ethene, we multiply all the amounts by 50.

Hence we have

$$C_2H_4 + HCl \longrightarrow C_2H_5Cl$$

$$50 \times 1 = 50\text{ g} \qquad 50 \times 1.3 = 65\text{ g} \qquad 50 \times 2.3 = 115\text{ g}$$

Maximum mass of chloroethane = 115 g

3 Lead (II) oxide can be reduced by carbon monoxide in a blast furnace to produce lead and carbon dioxide.

a) Write a balanced equation for this reaction.

b) 1 tonne (i.e. 1000 kg) is to be produced by this method.

What is the maximum theoretical mass in kg of lead(II) oxide that would be needed?

Answer

3 a) $PbO + CO \longrightarrow Pb + CO_2$

b) $PbO + CO \longrightarrow Pb + CO_2$

Pointer
These values are the molar masses of PbO and Pb looked up in the periodic table.

Pointer
We need to find what mass of PbO is needed to produce 1 g of Pb. Notice the way we divide through by 207.2 to make the mass of Pb equal to 1 g. The units can now be changed from g to kg. You can then multiply through by 1000 so that the mass of PbO needed to produce 1000 kg of Pb can be found.

PbO	Pb
1 mole	1 mole
223.2 g	207.2 g
$\frac{223.2}{207.2}$ g	$\frac{207.2}{207.2}$ g
= 1.0772 g	= 1 g
1.0772 kg	1 kg
1077.2kg	1000 kg

Maximum theoretical mass of PbO needed = 1077 kg

7.3 When one of the reactants limits the reaction

In some reactions the amounts of reactants are not stoichiometric. This means that one or more of the reactants are in excess. The reactant that is not in excess will limit the amount of product or products produced.

For example in the reaction:

$$MgO(s) + 2HNO_3(aq) \longrightarrow Mg(NO_3)_2(aq) + H_2O(l)$$

If reaction was stoichiometric:

$$MgO(s) + 2HNO_3(aq) \longrightarrow Mg(NO_3)_2(aq) + H_2O(l)$$

1 mole   2 moles   1 mole   1 mole

However, if the nitric acid was in excess and we had 6 moles of it but only 2 moles of MgO(s), the MgO is not in excess and would therefore be completely used up in the reaction. It would limit the amount of products formed and is called the limiting reactant.

The equation with the masses in moles for this reaction would now be:

$$MgO(s) + 2HNO_3(aq) \longrightarrow Mg(NO_3)_2(aq) + H_2O(l)$$

2 moles   4 moles   2 moles   2 moles

» Pointer

Notice that although there are 6 moles of nitric acid, we can only use up double the number of moles of MgO according to the stoichiometry of the equation.

Examples

1 When heated together, iron and sulfur form iron(II) sulfide. If 10.0 g of iron are heated with 10.0 g of sulfur, one of the reactants is in excess. Determine which of the reactants is in excess and determine the mass of iron(II) sulfide produced.

Answer

1 First we need to work out the number of moles of each reactant using the equation $n = \frac{m}{M}$

For Fe, number of moles in 10 g $= \frac{m}{M} = \frac{10.0}{55.8} = 0.1792$

For S, number of moles in 10 g $= \frac{m}{M} = \frac{10.0}{32.1} = 0.3115$

The equation for the reaction is $Fe + S \longrightarrow FeS$

As the ratio of Fe to S atoms is 1:1, the smaller amount of moles will limit the reaction.

» Pointer

The ratio is obtained using the stoichiometry of the equation.

There are 0.1792 moles of Fe which will only need 0.1792 moles of S to react completely.

Hence, the sulfur is in excess.

Now 0.1792 moles of FeS are produced. We need to find how much this is in grams.

Now rearranging $n = \frac{m}{M}$ to make m the subject of the equation, we obtain

$m = Mn = 87.9 \times 0.1792 = 15.75$ g (4 s.f.)

Hence 15.75 g of FeS are produced.

2 Boron trichloride (BCl_3) is prepared using the following reaction:

$$B_2O_3(s) + 3C(s) + 3\ Cl_2(g) \longrightarrow 2BCl_3(g) + 3CO(g)$$

A sample of 50.0 g of B_2O_3 was reacted with 10.0 g of C and 100.0 g of Cl_2.

a) Calculate the number of moles in 50.0 g of B_2O_3.

b) Determine which of the reactants are in excess.

c) Calculate the maximum number of moles of BCl_3 that can be produced from the sample.

Answer

2 a) Number of moles of B_2O_3, $n = \frac{m}{M} = \frac{50.0}{(2 \times 10.8 + 3 \times 16.0)} = 0.718$ mol

b) Number of moles of C, $n = \frac{m}{M} = \frac{10.0}{12.0} = 0.833$ mol

Number of moles of Cl_2, $n = \frac{m}{M} = \frac{100.0}{(35.5 \times 2)} = 1.41$ mol

According to the stoichiometry, we have

$B_2O_3(s)$	+	$3C(s)$	+	$3\ Cl_2\ (g)$	$\longrightarrow$	$2BCl_3(g)$	+	$3CO(g)$
1 mole		3 moles		3 moles		2 moles		3 moles
0.718		3 × 0.718		3 × 0.718				
0.718 moles		2.154 moles		2.154 moles				

Now there are only 0.833 mol of C and 1.41 mol of Cl_2 (note that according to the above 2.154 moles of these are needed to react with 0.718 moles of B_2O_3). This means that the B_2O_3 is in excess.

c) The substance with the smallest number of moles will limit the reaction. In this reaction this is the 0.833 mol of C.

Hence we can write the amounts for the reaction like this:

$B_2O_3(s)$	+	$3C(s)$	+	$3\ Cl_2\ (g) \longrightarrow$	$2BCl_3(g) + 3CO(g)$
$\frac{0.833}{3} = 0.278$ moles		0.833 moles		0.833 moles	0.278 × 2 = 0.556 moles

» *Pointer*

From the stoichiometry of the equation there is double the number of moles of BCl_3 compared to B_2O_3. Hence the number of moles of BCl_3 = 0.278 × 2 = 0.556.

Test yourself 7

1. Balance each of the following equations:
   a) $Mg + ...HCl \longrightarrow MgCl_2 + H_2$
   b) $...Na + ...H_2O \longrightarrow ...NaOH + H_2$
   c) $N_2 + ...H_2 \longrightarrow ...NH_3$
   d) $Cl_2 + ...KBr \longrightarrow ...KCl + Br_2$
   e) $...C + ...H_2 \longrightarrow C_4H_{10}$
   f) $...PbS + ...O_2 \longrightarrow ...PbO + ...SO_2$
   g) $BCl_3 + ...H_2O \longrightarrow H_3BO_3 + ...HCl$
   h) $CaCO_3 + ...HCl \longrightarrow CaCl_2 + CO_2 + H_2O$
   i) $...NaOH + CO_2 \longrightarrow Na_2CO_3 + H_2O$
   j) $...NaOH + H_2SO_4 \longrightarrow Na_2SO_4 + ...H_2O$

2. Barium sulfate, used in medicine for barium meals, can be prepared by reacting barium hydroxide with sulfuric acid.
   a) Write a balanced equation for this reaction.
   b) Calculate the maximum mass of barium sulfate that could be made using 2.0 g of barium hydroxide.

3. Ammonia, NH_3, is reacted with nitric acid to produce ammonium nitrate NH_4NO_3

$$NH_3 + HNO_3 \longrightarrow NH_4NO_3$$

   a) State the molar masses of
      i) Ammonia .... g mol^{-1}
      ii) Ammonium nitrate .... g mol^{-1}
   b) Calculate the maximum mass of ammonium nitrate, in tonnes, that can be made from 34.06 tonnes of ammonia.

4. The fertiliser ammonium sulfate is produced by reacting ammonia with sulfuric acid and the equation for the reaction is shown here:

$$2NH_3 + H_2SO_4 \longrightarrow (NH_4)_2SO_4$$

   a) State the molar masses of:
      i) Ammonia
      ii) Ammonium sulfate
   b) If 20 tonnes of ammonia are used, calculate the maximum mass in tonnes of ammonium sulfate that could be produced.

5. One way to prepare ethanol is by the fermentation of glucose:

$$C_6H_{12}O_6 \longrightarrow 2C_2H_5OH + 2CO_2$$

   Calculate the minimum mass of glucose required to give 230 g of ethanol.

Chapter 8 Radioactive decay and half-life

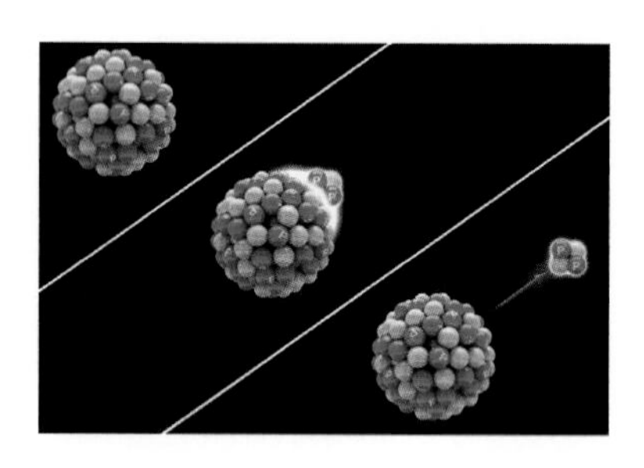

8.1 Radioactive decay

Unstable nuclei can become more stable by the emission of radiation. When an α or β particle is emitted, the nucleus changes into the nucleus of a different element because the proton number changes. The number of nuclei of the original element will decrease as will the mass, and the original element is said to undergo radioactive decay.

8.2 Half-life

The half-life, $T_{\frac{1}{2}}$, is the time taken for the mass of the original isotope to fall to half of its original value.

For example if 50 g of C-14 (a radioactive isotope of carbon) decayed and its mass reduced to 25 g in 5730 years, then the half-life of C-14 is 5730 years.

Half-life graph

The graph below shows the decay of a radioactive isotope of strontium called strontium-90.

The graph shows that the initial mass of 10 g of strontium-90 halves to 5 g in 29 years so 29 years is the half-life of the isotope.

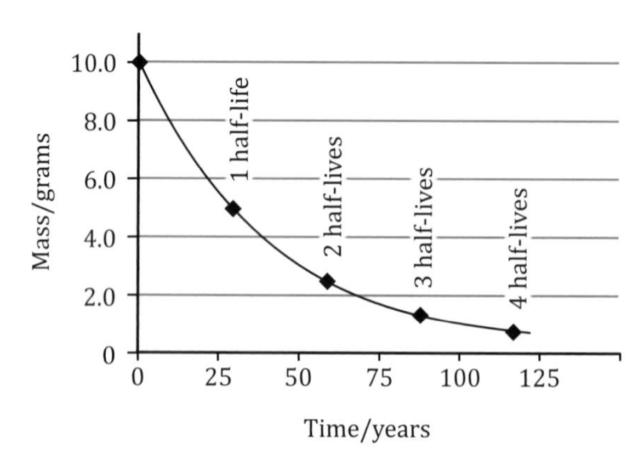

» Pointer

First find the number of half-lives in the time being considered. This is done by dividing the time being considered by the half-life of the isotope.

Examples

1 Tritium, which is an isotope of hydrogen, is a radioactive gas with a half-life of 12.5 years. A sample of tritium has a mass of 0.844 g.

Calculate the mass of tritium remaining after 50 years.

Answer

1 Number of half-lives in 50 years $= \dfrac{50}{12.5}$

$= 4$

Mass remaining after 50 years $= \frac{1}{2} \times \frac{1}{2} \times \frac{1}{2} \times \frac{1}{2} \times 0.844$

$= 0.053$ g

2 The radioactive isotope Potassium-40 decays by β-emission and has a half-life of 1.24×10^9 years.

Calculate how long it would take for a sample of the isotope to decay to $\frac{1}{16}$th of its original activity.

Answer

2 After 1 half-life the sample would have $\frac{1}{2}$ of its original activity.

After 2 half-lives the sample would have $\frac{1}{2} \times \frac{1}{2} = \frac{1}{4}$ of its original activity.

After 3 half-lives the sample would have $\frac{1}{2} \times \frac{1}{2} \times \frac{1}{2} = \frac{1}{8}$ of its original activity.

After 4 half-lives the sample would have $\frac{1}{2} \times \frac{1}{2} \times \frac{1}{2} \times \frac{1}{2} = \frac{1}{16}$ of its original activity.

4 half-lives $= 4 \times 1.24 \times 10^9 = 4.96 \times 10^9$ years

» Pointer

Note that after each half-life the mass will halve. There are 4 half-lives so we need to half the original mass four times.

» Pointer

You need to find how many times you halve the activity to give an activity of 1/16 of the original activity.

Test yourself 8

1. One of the isotopes of lead ^{212}Pb is radioactive and decays with a half-life of 10.6 hours. Calculate the amount of a sample of 56 mg of ^{212}Pb which remains after a time of 31.8 hours.
2. The radioactive isotope of carbon ^{14}C has a half-life of 5730 years. What fraction of the ^{14}C in a sample of the isotope would remain after a period of 11 460 years?
3. The half-life of caesium-137 is 30 years. Of an initial 100 g sample, how much caesium-137 will remain after 90 years?

Chapter 9 Percentage yield and atom economy

9.1 Percentage yield

In chemical reactions, reactants react to produce products.

The percentage yield is the amount of product actually made as a percentage of the amount the reaction should theoretically produce.

$$\text{General equation: Percentage yield} = \frac{\text{Actual yield}}{\text{Theoretical yield}} \times 100$$

Example

When limestone (calcium carbonate) is heated strongly, quicklime (calcium oxide) is produced along with the gas carbon dioxide.

The reaction is as follows:

$$CaCO_3 \longrightarrow CaO + CO_2$$

M_r values 100 56 44

Suppose this reaction is carried out industrially. The theory would predict that 100 tonnes of calcium carbonate would produce 56 tonnes of calcium oxide.

When this reaction actually takes place it is found that the yield of calcium oxide is only 48 tonnes.

$$\text{Hence the percentage yield} = \frac{\text{Actual yield}}{\text{Theoretical yield}} \times 100$$

$$= \frac{48}{56} \times 100$$

$$= 85.7\%$$

9.2 Percentage atom economy

Percentage atom economy is the mass of the desired product (i.e. the product you want to make) as a percentage of all the products made.

$$\text{General equation: Percentage atom economy} = \frac{\text{molar mass of desired product}}{\text{molar mass of all the products}} \times 100$$

For example, in the production of calcium oxide (slaked lime) from calcium carbonate (limestone) the reaction is as follows:

$$CaCO_3 \longrightarrow CaO + CO_2$$

M_r values 100 56 44

$$\text{Percentage atom economy} = \frac{\text{molar mass of desired product}}{\text{molar mass of all the products}} \times 100$$

$$= \frac{56}{(56 + 44)} \times 100$$

$$= 56\%$$

Test yourself 9

1. The following reaction occurs in the blast furnace when iron (III) oxide is reduced by carbon monoxide to produce iron:

$$Fe_2O_3 + 3CO \longrightarrow 2Fe + 3CO_2$$

M_r values 160 84 112 132

Calculate the % atom economy for this reaction.

2. The following reaction is used to convert ethene into chloroethane C_2H_5Cl by the addition of hydrogen chloride HCl:

$$C_2H_4 + HCl \longrightarrow C_2H_5Cl$$

M_r values 28.0 36.5 64.5

(a) In the above reaction, calculate the maximum theoretical mass of chloroethane produced when 70.0 g of ethene is used.

(b) An experiment was performed to produce chloroethane using the above reaction and it was found that when 70.0 g of ethene was used only 119 g of chloroethane was produced. Calculate the percentage yield in this experiment.

(c) Calculate the % atom economy for the reaction shown.

3. The industrial manufacture of ethanol is by the hydration of ethene according to the following reaction:

$$C_2H_4 + H_2O \longrightarrow C_2H_5OH$$

M_r values 28.0 46.0

In a typical process, 28 tonnes of ethene produces 43.7 tonnes of ethanol. Calculate the percentage yield of ethanol in this process.

Chapter 10 Determining relative atomic mass from relative or percentage abundance

10.1 Relative abundance of isotopes

The abundance of an isotope is the amount of the isotope present in a sample of different isotopes of the same element.

The relative abundance of an isotope is the abundance of the isotope divided by the total abundance of all the isotopes in the sample. Relative abundance has no units. Alternatively, relative abundances can be expressed as a percentage.

Example

An element has four isotopes and these appear in its mass spectrum as four peaks. The sample of this element has the following relative abundance as shown in this table.

m/z	64	66	67	68
Relative abundance	12	8	1	6

To determine the relative atomic mass of the element you need to find a weighted average (i.e. an average based on the relative abundance). Suppose we have 27 atoms (i.e. the relative abundance added together 12 + 8 + 1 + 6 = 27), 12 of these atoms would have m/z of 64, 8 would have m/z of 66 and so on.

The total mass of all 27 atoms can we worked out by multiplying the m/z value for each isotope by its relative abundance like this:

Total mass of 27 atoms $= (64 \times 12) + (66 \times 8) + (67 \times 1) + (68 \times 6)$

$= 1771$

So the average mass of one atom $= \frac{1771}{27}$

Relative atomic mass $= 65.6$ (3 s.f.)

10.2 Percentage abundance

Here the abundance of each isotope is expressed as a percentage, which means that the abundance of all the isotopes in a sample will add up to 100. Again a weighted average is used by multiplying each mass by the percentage and then adding to give the total mass.

Examples

1 A naturally occurring sample of potassium gave the following results for its mass spectrum:

Isotope	% abundance
^{39}K	93.26
^{40}K	0.012
^{41}K	6.730

The results in the table can be used to calculate the relative atomic mass of the potassium sample.

Calculate the relative atomic mass of the potassium sample, giving your answer correct to **four** significant figures.

Answer

1 Relative atomic mass, $A_r = \left(\frac{93.26}{100} \times 39\right) + \left(\frac{0.012}{100} \times 40\right) + \left(\frac{6.730}{100} \times 41\right)$

$= 39.14$ (four significant figures)

Always ask yourself if the question specifies that the answer be given to a certain number of significant figures or decimal places.

Note that the isotope with the higher % abundance has a greater weighting, so the A_r of the sample will be nearer to this value.

2

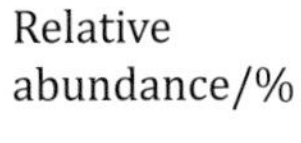

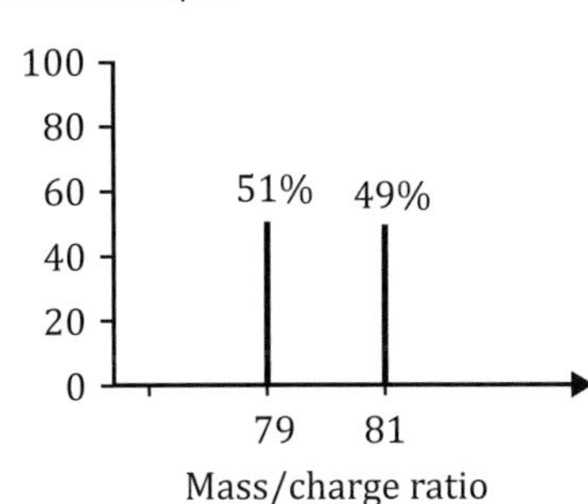

A sample of copper contains two isotopes. The percentage abundance of each isotope along with its relative atomic mass is shown in the graph above. Calculate the relative atomic mass of the sample of copper, giving your answer to three significant figures.

Answer

2 $A_r = \left(\frac{51}{100} \times 79\right) + \left(\frac{49}{100} \times 81\right) = 79.98 = 80.0$ (3 s.f.)

Test yourself 10

1. The mass spectrum of a sample of boron gave the following results:

Isotope	% abundance
^{10}B	18.70
^{11}B	81.30

Use the results in the table to calculate the relative atomic mass of the boron sample to four significant figures.

2. Copper consists mainly of two isotopes, ^{63}Cu and ^{65}Cu. In a typical sample of copper 72.5% of the atoms are ^{63}Cu and 27.5% are ^{65}Cu. Calculate the relative atomic mass (A_r) for this sample of copper. Give your answer correct to three significant figures.

3. The mass spectrum of a sample of krypton is summarised in the following table:

m/z	Relative abundance
82	3.0
83	2.0
84	10.0
86	3.0

Use the data in the table to calculate a value for the relative atomic mass of this sample of krypton, giving your answer correct to one decimal place.

Chapter 11 Concentrations

When a substance (called the solute) dissolves in a liquid (called the solvent), the amount of solute dissolved in a certain volume of solvent is called the concentration of the resulting solution.

Concentration can be measured in the following different units:

mol dm^{-3}

mol l^{-1}

ppm (parts per million).

Pointer

Spend some time making sure you can change volumes from one unit to another. Many students make mistakes in exam questions for failing to make the conversions correctly.

11.1 Converting volumes

1 dm = 10 cm, so 1 dm^3 = 10 × 10 × 10 cm^3 = 1000 cm^3

Also, 1000 cm^3 = 1 litre (i.e abbreviated to 1 l)

To convert from cm^3 to dm^3 you divide by 1000.

So, for example 250 $cm^3 = \frac{250}{1000}$ dm^3 = 0.25 dm^3.

To convert from dm^3 to cm^3 you multiply by 1000.

So, for example 0.04 dm^3 = 0.04 × 1000 = 40 cm^3.

As 1000 cm^3 = 1 dm^3 = 1 l, one litre and one dm^3 are the same volume and can be used interchangeably.

Working out concentrations in mol dm^{-3}

If the volume V is in dm^3 and n is the number of moles, then concentration, c in mol dm^{-3} is given by:

$$c = \frac{n}{V} \text{ mol dm}^{-3}$$

If the volume V is in cm^3 and n is the number of moles, then concentration, c in mol dm^{-3} is given by:

$$c = \frac{1000n}{V} \text{ mol dm}^{-3}$$

Pointer

Remember that dm^3 and l can be used interchangeably.

Examples

1 Calculate the concentration of 0.01 moles of sodium hydroxide (NaOH) dissolved in water to make up 50 cm^3 of solution.

Answer

1 50 $cm^3 = \frac{50}{1000}$ dm^3 = 0.05 dm^3

Pointer

You could also have used the alternative formula of $c = \frac{1000n}{V}$ where V is the volume in cm^3.

Now, $c = \frac{n}{V} = \frac{0.01}{0.05} = 0.2 \text{ mol dm}^{-3}$

2 A solution of NaCl has a concentration of 0.5 mol dm^{-3}. Calculate the number of moles of NaCl dissolved in 25 cm^3 of this solution.

Answer

2 Now, $c = \frac{n}{V}$, so $n = cV = 0.5 \times \frac{25}{1000} = 0.0125 \text{ mol}$

3 Carbon dioxide is dissolved in water to make fizzy drinks. The solubility of carbon dioxide, $M_r = 44$, in water at 25°C and atmospheric pressure is

0.145 g/100 g H_2O.

Calculate the concentration of the resulting solution in mol dm^{-3}.

Answer

3 0.145 g/100 g H_2O is the same concentration as 1.45 g/1000 g H_2O

Now 1000 g of water has a volume of 1000 cm^3 = 1 dm^3

Hence 1.45 g of CO_2 can be dissolved in 1 dm^3 of H_2O.

Number of moles, n, in 1.45 g of CO_2 is given by

$n = \frac{m}{M} = \frac{1.45}{44} = 0.0330$ moles

Using $c = \frac{n}{V}$

we obtain $c = \frac{0.0330}{1} = 0.0330 \text{ mol dm}^{-3}$

4 In August 2013, the average concentration of CO_2 in the atmosphere was 394 ppm. What percentage is this?

Answer

4 In one million parts of air there would be 394 parts of CO_2.

Fraction of CO_2 in the air is $\frac{394}{1\,000\,000}$

Percentage of CO_2 in the air $= \frac{394}{1\,000\,000} \times 100 = 0.0394\%$

5 Silver chloride (AgCl) is almost insoluble in water. The solubility of AgCl in water has been estimated to be 0.2 mg in 125 cm^3 of water at 20°C.

a) Calculate the molar mass of silver chloride.

b) Calculate the number of moles of silver chloride in 0.2 mg.

c) Calculate the solubility of silver chloride in water in moles per litre.

Answer

5 a) Molar mass of AgCl, $M = 107.9 + 35.5$

$= 143.4 \text{ g mol}^{-1}$

b) Number of moles, $n = \frac{m}{M} = \frac{0.2 \times 10^{-3}}{143.4} = 1.39 \times 10^{-6} \text{ mol}$

Pointer
The volume is in cm^3 and needs to be converted to dm^3 by dividing by 1000.

Pointer
You need to remember that 1 g of water has a volume of 1 cm^3.

Pointer
Note that in the formula $c = n/V$, the volume must be in dm^3.

Pointer
Remember that to turn a fraction into a percentage you multiply by 100.

Pointer
These two values are the A_r of Ag and Cl obtained from the periodic table.

c) Now there are 1.39×10^{-6} mol dissolved in 125 cm^3

Since $8 \times 125 = 1000\ cm^3 = 1$ litre, there will be

$8 \times 1.39 \times 10^{-6}\ mol\ l^{-1} = 1.112 \times 10^{-5}\ mol\ l^{-1}$

Hence, concentration of AgCl = $1.11 \times 10^{-5}\ mol\ l^{-1}$

11.2 Concentrations in parts per million (ppm)

The concentration of a substance in solution can be expressed in parts per million (ppm). This means the number of parts of solute that are dissolved in one million parts of the solvent.

Examples

1 1 kg of a solution contains 0.100 mol of calcium ions, Ca^{2+}. Calculate the concentration of the calcium ions by mass in parts per million (ppm). (Assume the relative atomic mass of calcium is 40.)

Answer

1 Using $n = \frac{m}{M}$ and rearranging for m gives $m = Mn$.

$m = Mn = 40 \times 0.100 = 4$ g

This means there are 4 g of Ca^{2+} ions in 1000 g of solution.

There will be 4×1000 of Ca^{2+} ions in $1000 \times 1000 = 1\,000\,000$ g of solution.

Hence the concentration of Ca^{2+} ions in solution is 4000 ppm.

2 A sample of blood plasma contained 20 parts per million (ppm) of magnesium.

Calculate the mass of magnesium in grams that would be present in 100 g of the sample of blood plasma.

Answer

2 1 000 000 parts of blood plasma contains 20 parts of Mg

$\frac{1\,000\,000}{10\,000}$ parts of blood plasma contains $\frac{20}{10\,000}$ parts of Mg

100 parts of blood plasma contains 0.002 parts of Mg

Hence 100 g of blood plasma contains 0.002 g of Mg

Pointer

As the question is concerned with masses, we need to find the mass of Ca^{2+} in g in 0.100 mol.

Pointer

Here we have a ratio of 4:1000 which needs to be turned into a ratio where the second number is one million. We can do this by multiplying both numbers in the ratio by 1000.

Pointer

To turn 1 000 000 parts of blood plasma to 100 parts you divide by 10 000. You then have to divide the 20 parts of Mg by 10 000 to keep the ratio the same. You can then change the units to grams.

11.3 Concentrations used to calculate masses

Concentrations are often used in calculations to work out masses of products or reactants in reactions. If a concentration and volume of a solution are known then the number of moles in the solution can be found. This can then be used to determine the number of moles and hence the mass in grams for the other products and reactants in a reaction as the following example shows.

Example

Magnesium chloride can be prepared by reacting magnesium carbonate with dilute hydrochloric acid. The equation for this reaction is shown below:

$MgCO_3(s) + 2HCl(aq) \longrightarrow MgCl_2(aq) + CO_2(g) + H_2O(l)$

To make crystals of hydrated magnesium chloride, $MgCl_2.6H_2O$, magnesium carbonate was added to 25 cm^3 of hydrochloric acid with concentration 2.0 mol dm^{-3}. The molar mass of magnesium carbonate is 84.3 g mol^{-1}.

a) Calculate the number of moles of acid used in the reaction.

b) What mass of magnesium carbonate, in grams, reacts with this amount of acid?

» Pointer
Here we use the volume and concentration of the acid to work out the number of moles of acid.

Answer

a) $V = \frac{25}{1000} = 0.025\ dm^3$

Using $c = \frac{n}{V}$ and rearranging for n gives $n = cV = 2.0 \times 0.025 = 0.05$ mol

b) According to the stoichiometry of the equation:

$MgCO_3(s) + 2HCl(aq) \longrightarrow MgCl_2(aq) + CO_2(g) + H_2O(l)$

$\frac{0.05}{2}$ mol   0.05 mol

Hence 0.025 mol of $MgCO_3$ are needed

Now $n = \frac{m}{M}$, so $m = Mn = (24.3 + 12.0 + 3 \times 16.0) \times 0.025 = 2.11$ g

Test yourself 11

1. Calculate the number of moles of copper(II) sulfate in 50.0 cm^3 of a 0.500 mol dm^{-3} solution.

2. Calculate the concentration, in mol dm^{-3}, of a solution formed when 10 g of NaOH is dissolved in water to make up 400 cm^3 of solution.

3. Ethanol (C_2H_5OH) is the alcoholic component in alcoholic drinks.

   The legal limit for ethanol in a driver's blood is 80 mg per 100 cm^3.

   Find this concentration of ethanol in mol dm^{-3}.

Chapter 12 The ideal gas equation

12.1 The ideal gas equation

The ideal gas equation is an equation that is obeyed by an ideal gas, although no real gas obeys it exactly in reality.

The ideal gas equation connects the number of moles of gas, pressure, volume, temperature and a constant called R together.

The ideal gas equation states:

$$pV = nRT$$

where

p is the pressure of the gas in pascals (Pa)

V is the volume of the gas in m^3

n is the amount of gas in mol

R is a constant called the gas constant which has a value of $8.31\ J\ K^{-1}\ mol^{-1}$

T is the temperature in kelvin (K)

You must be able to rearrange this equation and substitute numbers into it. You must also be able to convert volumes which are not in the units of m^3 into m^3 so they can be substituted into this formula. If you are unsure of this take a look back at the maths section on page 33.

Pointer
Be careful with volumes of gases. Make sure you can convert from volumes in cm^3 to volumes in m^3. Note that $1\ cm^3$ is equal to $1 \times 10^{-6}\ m^3$.

Pointer
Temperatures in K can be obtained by adding 273 to the temperature in °C.

Examples

1 When heated, limestone ($CaCO_3$) decomposes to produce quicklime (CaO) and carbon dioxide (CO_2) according to the following equation.

$$CaCO_{3\,(s)} \longrightarrow CaO_{\,(s)} + CO_{2\,(g)}$$

A sample of limestone was decomposed completely.

The carbon dioxide produced occupied a volume of $220\ cm^3$ at a pressure of 100 kPa and a temperature of 28°C.

(The gas constant $R = 8.31\ J\ K^{-1}\ mol^{-1}$)

Calculate the number of moles of carbon dioxide produced.

Pointer
As the question includes the gas constant, R, it means that the question involves the use of the ideal gas equation.

Answer

1 $V = 220\ cm^3 = 220 \times 1 \times 10^{-6}\ m^3 = 2.20 \times 10^{-4}\ m^3$

$T = (28 + 273) = 301\ K$

Using $pV = nRT$

Pointer
Remember that when entering volumes into the ideal gas equation, the volumes must be in m^3.

» Pointer
Divide both sides of the original formula by RT to make n the subject.

» Pointer
Note that 100 kPa = 100×10^3 Pa.

Making n the subject of the formula gives

$$n = \frac{pV}{RT}$$

$$\text{Hence, } n = \frac{100 \times 10^3 \times 2.20 \times 10^{-4}}{8.31 \times 301}$$

$$= 8.80 \times 10^{-3} \text{ moles (3 s.f.)}$$

2 Calcium nitrate decomposes on heating to form calcium oxide, nitrogen dioxide and oxygen according to the following equation.

$2Ca(NO_3)_2(s) \longrightarrow 2CaO(s) + 4NO_2(g) + O_2(g)$

A sample of calcium nitrate was decomposed completely.

The gases produced occupied a volume of 2.75×10^{-3} m^3 at a pressure of 100 kPa and a temperature of 28°C.

(The gas constant $R = 8.31$ J K^{-1} mol^{-1})

a) Calculate the total amount in moles of gases produced.

b) Hence calculate the amount in moles of nitrogen dioxide produced.

Answer

2 a) Using $pV = nRT$

Making n the subject of the formula gives

» Pointer
Divide both sides of the original formula by RT to make n the subject.

» Pointer
Note $T = 28 + 273 = 301$ K

$$n = \frac{pV}{RT}$$

$$n = \frac{100 \times 10^3 \times 2.75 \times 10^{-3}}{8.31 \times 301}$$

$$= 0.1099$$

$$= 0.110 \text{ mol (3 s.f.)}$$

b) The ratio of NO_2 to O_2 is 4:1.

Hence $\frac{4}{5}$ of the total moles of gases are NO_2.

So, the number of moles of $NO_2 = \frac{4}{5} \times 0.1099 = 0.088$ mol (3 s.f.).

Test yourself 12

1. In an experiment producing nitrogen monoxide gas, 638 cm^3 of the gas was produced at 101 kPa and 25°C. Calculate the number of moles of nitrogen gas in this volume and under these conditions.

   (The gas constant R = 8.31 J K^{-1} mol^{-1})

2. Calculate the volume in m^3 of 2.5×10^{-2} moles of oxygen gas at a pressure of 150 kPa and a temperature of 50°C.

   (The gas constant R = 8.31 J K^{-1} mol^{-1})

3. A cylinder of carbon dioxide contains 10 moles of gas. If the volume of the cylinder is 2000 cm^3 and the temperature of the gas is 25°C, calculate the pressure of the gas giving your answer in kPa correct to three significant figures.

   (The gas constant R = 8.31 J K^{-1} mol^{-1})

4. Potassium nitrate decomposes on strong heating to form oxygen gas and solid potassium oxide.

   (a) A 1.00 g sample of potassium nitrate was heated strongly. Calculate the number of moles of potassium nitrate in 1.00 g.

   (b) The oxygen gas produced by this decomposition occupied a volume of 1.22×10^{-4} m^3 at a temperature of 298 K and a pressure of 100 kPa. Using the ideal gas equation, determine the number of moles of oxygen produced during the decomposition.

Chapter 13 Acid–base titrations

13.1 Acid–base titrations

A solution of acid of known concentration, called a standard solution, can be used to find the concentration of an alkali using an acid–base titration. In such a titration, a known volume of the alkali is transferred using a pipette to a conical flask and a few drops of indicator are added. The acid solution (the concentration of which is known) is added to a burette and then added to the alkali slowly (i.e. titrated) until a colour change occurs showing that the acid had just neutralised the alkali. The volume of acid added is found by subtracting an initial and final burette reading. A mean value for the volume is determined by performing the titration several times.

The following example shows how the results of a titration can be used to determine the concentration of alkali.

Example

1 A 25.0 cm^3 solution of sodium hydroxide required 20.0 cm^3 of hydrochloric acid of concentration of 0.100 mol dm^{-3} for neutralisation. Find the concentration of the solution of sodium hydroxide.

Answer

1 There are a number of steps to take:

Pointer
Note carefully the stoichiometry of the equation. Here the ratio of acid to base is 1:1 but this is not always the case.

① Write a balanced equation for the neutralisation.

$$NaOH + HCl \longrightarrow NaCl + H_2O$$

1 mole 1 mole

Notice that one mole of alkali is needed to neutralise one mole of acid.

② Work out the number of moles in 20.0 cm^3 of 0.100 mol dm^{-3} hydrochloric acid.

Use the formula $c = \frac{n}{V}\left(\text{i.e. concentration in mol dm}^{-3} = \frac{\text{number of moles in mol}}{\text{volume in dm}^3}\right)$

Pointer
Note that the volume needs to be changed from cm^3 to dm^3. To convert cm^3 to dm^3 you divide the volume by 1000.

Rearranging for n gives $n = cV$

Hence $n = cV = 0.100 \times \frac{20.0}{1000}$

$= 0.002$ moles

③ Use the stoichiometry of the equation to determine the number of moles of sodium hydroxide needed for neutralisation.

From the equation the ratio of number of moles of acid to number of moles of alkali is one to one.

However, there are 0.002 moles of acid so 0.002 moles of alkali are needed for neutralisation.

Volume of the alkali used = 25.0 cm^3 = $\frac{25.0}{1000}$ dm^3 = 0.025 dm^3

Concentration of NaOH, $c = \frac{n}{V} = \frac{0.002}{0.025}$ = 0.080 mol dm^{-3} (3 s.f.)

Examples

1 In an experiment, a student titrated 25.00 cm^3 of sodium hydroxide solution with hydrochloric acid, and obtained the following results for the burette readings:

	1	2	3	4
Initial burette reading/cm^3	1.75	3.75	12.55	20.00
Final burette reading/cm^3	18.85	20.55	30.05	37.20
Volume used/cm^3				

a) Complete the table to show the volume of acid used in each titration.

b) Calculate the mean volume of acid that should be used for further calculations.

Answer

1 a)

	1	2	3	4
Initial burette reading/cm^3	1.75	3.75	12.55	20.00
Final burette reading/cm^3	18.85	20.55	30.05	37.20
Volume used/cm^3	17.10	16.80	17.50	17.20

b) Mean volume used = $\frac{17.10 + 16.80 + 17.50 + 17.20}{4}$ = 17.15 cm^3

Pointer

Arithmetic means are covered in the maths section of this book on page 28.

2 Aqueous ammonia solution reacts with sulfuric acid according to the following equation:

$$2NH_3(aq) + H_2SO_4(aq) \longrightarrow (NH_4)_2SO_4(aq)$$

A solution of 500 cm^3 of NH_3(aq) was obtained and a student was asked to determine the number of moles of ammonia that was dissolved to make this solution.

The student used a 25 cm^3 sample of the ammonia solution which she then titrated with sulfuric acid solution. It was found that 35 cm^3 of 0.100 mol dm^{-3} sulfuric acid solution, reacted exactly with the 25 cm^3 of ammonia solution.

Calculate the number of moles of NH_3 in the original 500 cm^3 of solution.

Answer

2 $2NH_3(aq) + H_2SO_4(aq) \longrightarrow (NH_4)_2SO_4(aq)$

2 moles   1 mole

No of moles, n, of H_2SO_4 in 35 cm^3 of 0.100 mol dm^{-3} is given by

$n = cV$

$= 0.100 \times \frac{35}{1000}$

$= 0.0035$ mol

Pointer

Determine the stoichiometry of the equation to determine the ratio of acid to alkali which is 1:2 in this case.

Pointer

Note that from the equation, the ratio of H_2SO_4 to NH_3 is 1:2.

0.0035 mol of H_2SO_4 neutralise 2×0.0035

$= 0.0070$ mol of NH_3

Pointer
The volume has been converted from cm^3 to dm^3.

Concentration of $NH_3(aq) = \frac{n}{V} = \frac{0.0070}{0.025} = 0.28 \text{ mol dm}^{-3}$

There are 0.28 mol in 1 dm^3.

So there will be 0.14 mol in 500 cm^3.

Pointer
Note that 500 cm^3 = 0.5 dm^3

3 Sulfuric acid is neutralised by aqueous sodium hydroxide according to the following equation:

$$H_2SO_4(aq) + 2NaOH(aq) \longrightarrow Na_2SO_4(aq) + 2H_2O(l)$$

A mass *m* g of solid sodium hydroxide is dissolved in water and the solution is made up to a volume of 250 cm^3.

It was found that 25.0 cm^3 of this solution was neutralised by 28.0 cm^3 of 0.100 mol dm^{-3} dilute sulfuric acid.

a) Calculate the number of moles of sulfuric acid used for neutralisation.

b) Determine the number of moles of sodium hydroxide in 25.0 cm^3.

c) Determine the number of moles of sodium hydroxide in 250 cm^3 of solution and hence calculate the mass *m* in g of sodium hydroxide that was dissolved.

Answer

3 a) No of moles, *n*, of H_2SO_4 in 28 cm^3 of 0.100 mol dm^{-3} is given by

$$n = cV$$
$$= 0.100 \times \frac{28}{1000}$$
$$= 0.0028 \text{ mol}$$

b) $H_2SO_4(aq) + 2NaOH(aq) \longrightarrow Na_2SO_4(aq) + 2H_2O(l)$

1 mole 2 moles

Number of moles of NaOH in 25.0 $cm^3 = 2 \times 0.0028 = 0.0056$ mol

Pointer
Notice that the ratio of moles of H_2SO_4 to moles of NaOH is 1:2.

c) Number of moles of NaOH in 250 cm^3 of solution $= 10 \times 0.0056$

$= 0.056$ mol

M for NaOH = $(23.0 + 16.0 + 1.0) = 40 \text{ g mol}^{-1}$

Now, $n = \frac{m}{M}$. Rearranging for *m* gives $m = nM = 0.056 \times 40 = 2.24$ g

Hence, $m = 2.24$ g

Test yourself 13

1. 30 cm^3 of a solution of potassium hydroxide (KOH) of concentration to be determined, required 45 cm^3 of 0.100 mol dm^{-3} of sulfuric acid for neutralisation. Calculate the concentration of the alkali.

2. A 0.02 mol sample of magnesium oxide reacts with hydrochloric acid according to the following equation:

   $$MgO + 2HCl \longrightarrow MgCl_2 + H_2O$$

   (a) Calculate the number of moles of acid that would be needed to react completely with 0.02 mol sample of magnesium oxide.

   (b) The 0.02 mol sample of magnesium oxide required 30.0 cm^3 of hydrochloric acid for complete reaction. Using this and your answer to part (a) calculate the concentration in mol dm^{-3} of the acid.

3. Calculate the volume of 0.08 mol dm^{-3} sodium hydroxide that would be needed just to neutralise 25.0 cm^3 of 0.10 mol dm^{-3} of sulfuric acid. You should give the volume in cm^3.

Chapter 14 Determining molecular and empirical formulae

Molecular formula

The molecular formula shows the number of atoms of each element present in one molecule of a compound. For example, the molecular formula of ethane is C_2H_6, showing that in one molecule there are 2 atoms of carbon and 6 atoms of hydrogen. The molecular formula is the formula we use in most chemical equations.

Empirical formula

The empirical formula of a compound is the simplest whole number ratio of the atoms of element present in the compound.

14.1 The difference between molecular and empirical formulae

To see the difference between molecular and empirical formulae, we can take ethene as an example.

The molecular formula for ethene is C_2H_4.

The empirical formula gives the simplest whole number ratio of atoms.

In C_2H_4 the simplest ratio of carbon to hydrogen atoms is 1:2 so the empirical formula is CH_2.

» Pointer
Note that in C_2H_4 the ratio of carbon to hydrogen atoms is 2:4 and this is simplified to 1:2.

In order to find the empirical formula from the percentage of each element in the compound, you first check that all the percentages add up to 100. If the compound is a hydrocarbon and you are told that it contains 64.4% of carbon, then the percentage of hydrogen is found by subtracting 64.4 from 100 to give 35.6%

You then divide each percentage by the A_r of the element.

You then divide each of the resulting numbers by the smallest number. In most cases you will get numbers which are integers or very near to integers which gives the empirical formula. This is best seen by looking at some examples.

Examples

1 A compound consisting of carbon, hydrogen and oxygen has a percentage composition by mass of C 40.0%, H 6.7%, O 53.3% and a relative molecular mass of 180.

a) Determine the empirical formula of this compound.

b) Determine the molecular formula of this compound.

Answer

1 a) First set out the percentages for each element in the compound as follows

C	H	O
40.0	6.7	53.3

Then divide each percentage by the molar mass of each element

C	H	O
$\frac{40.0}{12} = 3.333$	$\frac{6.7}{1.01} = 6.634$	$\frac{53.3}{16.0} = 3.331$

Divide each of these numbers by the smallest number (i.e. 3.331 in this case)

C	H	O
$\frac{3.333}{3.331} \approx 1$	$\frac{6.634}{3.331} \approx 2$	$\frac{3.331}{3.331} = 1$

Empirical formula is CH_2O

b) The empirical formula mass = (1 × 12) + (2 × 1.01) + (1 × 16.0)

= 30.02

The relative molecular mass = 180 ≈ 6 × empirical formula mass.

Hence molecular formula is $C_6H_{12}O_6$

2 An oxide of nitrogen contains 30.4% of nitrogen and 69.6% of oxygen, by mass. The relative molecular mass of the oxide is 92.

Calculate:

a) The empirical formula.

b) The molecular formula of this oxide.

Answer

2 a)

N	O
$\frac{30.4}{14.0} = 2.17$	$\frac{69.6}{16.0} = 4.35$
$\frac{2.17}{2.17} = 1$	$\frac{4.35}{2.17} \approx 2$

Hence the empirical formula is NO_2

b) M_r of NO_2 = 14.0 + (2 × 16.0) = 46

Now M_r of the oxide given in the question is 92.

This is 2 × M_r of empirical formula of the oxide.

Hence, molecular formula = N_2O_4

» Pointer
Divide each % of the element by its corresponding A_r.

» Pointer
Divide each by the smallest number i.e. 2.17 in this case.

» Pointer
Note that the symbol ≈ means is approximately equal to.

» Pointer
Remember to do the multiplication first.

» Pointer
The empirical formula needs to be multiplied by two to give the molecular formula of the oxide.

Test yourself 14

1. Benzene is a hydrocarbon compound of 92.3% carbon and 7.7% hydrogen by mass.
   a) Calculate the empirical formula of this compound.
   b) The relative molecular mass of benzene is 78.06. Give the molecular formula of this compound.

2. A compound consists of 25.5% sulfur and 36.5% sodium by mass, with the rest being oxygen. Determine the empirical formula of the compound.

3. Phosgene, a poisonous gas used in the First World War, is a compound consisting of the elements carbon, oxygen and chlorine. Its percentage composition by mass is as follows:

   C 12.1% O 16.2% Cl 71.7%

   a) Calculate the empirical formula of phosgene.
   b) What other information would be needed in order to determine the molecular formula of this compound?

4. Magnesium nitride can be produced by heating magnesium metal in a pure nitrogen atmosphere. Magnesium nitride contains 72.2% by mass of magnesium.

   Showing your working, calculate the empirical formula for magnesium nitride.

Chapter 15 Calculating the formula of a hydrated salt

Many crystalline salts contain water molecules and such salts are said to be hydrated. Hydrated salts contain water of crystallisation and the number of molecules of water is shown after the formula with a dot like this:

$$CuSO_4.5H_2O$$

Many hydrated salts can have their water of crystallisation driven off when heated strongly, leaving an anhydrous powder. For example, blue hydrated copper(II) sulfate crystals, $CuSO_4.5H_2O$, when heated strongly, form white anhydrous copper(II) sulfate powder $CuSO_4$.

15.1 Finding the formula of a hydrated salt

In order to determine the formula of a hydrated salt you need to take a known mass of the hydrated salt and drive off the water of crystallisation and then measure the mass of the anhydrous salt. Subtraction of the two masses gives the mass of the water of crystallisation in the sample. The following example shows how to then calculate the number of molecules of water in the formula.

Examples

1 Crystals of calcium nitrate contain water of crystallisation and its formula can be represented as $Ca(NO_3)_2.xH_2O$ where x is an integer.

A sample of 18.12 g of $Ca(NO_3)_2.xH_2O$ contained 5.52 g of water of crystallisation.

Using this information, calculate the value of x.

Answer

1 Mass of anhydrous calcium nitrate = 18.12 – 5.52 = 12.6 g

Number of moles of anhydrous calcium nitrate $= \frac{12.6}{(40.1 + (2 \times 14.0)+(6 \times 16.0)}$

$= 0.07678$

Number of moles of water $= \frac{m}{M} = \frac{5.52}{(2 \times 1.0 + 16.0)} = 0.3067$

$CaNO_3$	H_2O
0.07678	0.3067

Pointer

Write down the number of moles of the anhydrous salt and the the water.

$$\frac{0.07678}{0.07678} = 1 \qquad \frac{0.3067}{0.07678} \approx 4$$

Hence, $x = 4$ (as x has to be an integer)

Divide by the smallest number.

An integer is a whole number (which can be positive, negative or zero).

2 Hydrated calcium sulfate crystals, $CaSO_4.xH_2O$, have a relative formula mass of 172.2. Calculate the value of x.

Answer

2 Relative formula mass of $CaSO_4 = 40.1 + 32.1 + 4 \times 16 = 136.2$

Relative formula mass of $CaSO_4.xH_2O = 172.2$

In one mole the mass in g of the water = 172.2 − 136.2 = 36.0 g

Relative molecular mass of $H_2O = 2 \times 1 + 16 = 18$

No of water molecules in 36.0 g $= \frac{36.0}{18} = 2$

Hence, $x = 2$

3 a) Calculate the molar mass in g mol^{-1} of copper sulfate pentahydrate, $CuSO_4.5H_2O$.

b) Calculate the percentage of water by mass in the copper sulfate pentahydrate.

3 a) M of $CuSO_4.5H_2O = 63.5 + 32.1 + (4 \times 16.0) + (5 \times 18) = 249.6$ g mol^{-1}

b) Percentage of water by mass $= \frac{5 \times \text{molar mass of water}}{\text{molar mass of copper sulfate pentahydrate}} \times 100$

$= \frac{5 \times 18.0}{249.6} \times 100$

$= 36.1$ % (3 s.f.)

Test yourself 15

1. Hydrated copper(II) sulfate can be produced by allowing a solution of copper(II) sulfate to evaporate.

   Hydrated copper(II) sulfate has the empirical formula $CuSO_9H_{10}$.

   Write the formula of hydrated copper(II) sulfate to show its water of crystallisation.

2. A solution of sodium sulfate (Na_2SO_4) was allowed to evaporate and produce crystals with a formula $Na_2SO_4.\mathbf{x}H_2O$ and having a molar mass of 322.1 g mol^{-1}.
   (a) What term is given to the '$.\mathbf{x}H_2O$' part of the formula?
   (b) Use the molar mass of the crystals to calculate a value for **x**.

3. Sodium thiosulfate crystals have the formula $Na_2S_2O_3.5H_2O$. When heated strongly to remove the water of crystallisation, a white powder called anhydrous sodium thiosulfate forms.
   (a) Calculate the relative formula mass of $Na_2S_2O_3.5H_2O$
   (b) Calculate the mass in g of water produced when 25 g of hydrated sodium thiosulfate crystals are heated strongly.

Chapter 16 Enthalpy changes and calorimetry

16.1 Enthalpy changes ΔH

During chemical reactions there are energy changes and as a result, heat is usually either absorbed or given out. During reactions energy is used to break bonds and energy is released when new bonds are formed. It all depends on the differences between these two amounts of energy, whether energy is absorbed or released.

The amount of heat absorbed or released during the reaction is called the enthalpy change or ΔH for short.

When heat is evolved in a reaction (i.e. heat is released from the system) the sign of the enthalpy change is negative (i.e. ΔH has a negative sign) and the reaction is said to be **exothermic**.

When heat is absorbed in a reaction (i.e. heat is absorbed by the system) the sign of the enthalpy change is positive (i.e. ΔH has a positive sign) and the reaction is said to be **endothermic**.

In an exothermic reaction heat is given to the surroundings, resulting in an increase in temperature; and in an endothermic reaction, heat is absorbed from the surroundings, resulting in a decrease in temperature of the surroundings.

The amount of heat energy exchanged with the surroundings

The amount of heat energy (in J) exchanged with the surroundings is given the symbol Q and can be calculated if the following are known for the surroundings:

- the mass (m) in grams
- the specific heat capacity (c) in $\text{J g}^{-1}\text{ K}^{-1}$
- the temperature change ΔT (in either °C or K)

The equation connecting these quantities is:

$$Q = mc\Delta T$$

The specific heat capacity varies depending on the material of the surroundings and is a constant for a particular material. The value is usually given in the question.

The enthalpy change for a reaction

Enthalpy change, ΔH, for a reaction, is expressed in kJ mol^{-1}, the heat absorbed or evolved during the reaction. This can be used to increase or lower the temperature of the surroundings and this amount of heat can be determined by the above equation.

It should also be noted that the formula for Q shown above will give the heat energy in J exchanged with the surroundings. Enthalpy change is measured in kJ mol^{-1}. So to work out the enthalpy change in kJ mol^{-1} you would need to divide by the number of moles of substance providing the energy change and also to divide by 1000 to change the heat from J to kJ.

Hence, enthalpy change ΔH in kJ mol^{-1} is given by the formula

$$\Delta H = \frac{\text{Energy exchanged with surroundings in kJ}}{\text{Number of moles of substance}} = \frac{Q\ (\text{kJ})}{n\ (\text{mol})}$$

Enthalpy increases or decreases

If there is an enthalpy change during a reaction, then the enthalpy can either increase or decrease.

If the enthalpy decreases there is a negative enthalpy change (i.e. ΔH is negative) and the reaction gives out heat and is therefore exothermic.

If the enthalpy increases there is a positive enthalpy change (i.e. ΔH is positive) and the reaction takes in heat and is therefore endothermic.

This can be seen in the following enthalpy diagrams.

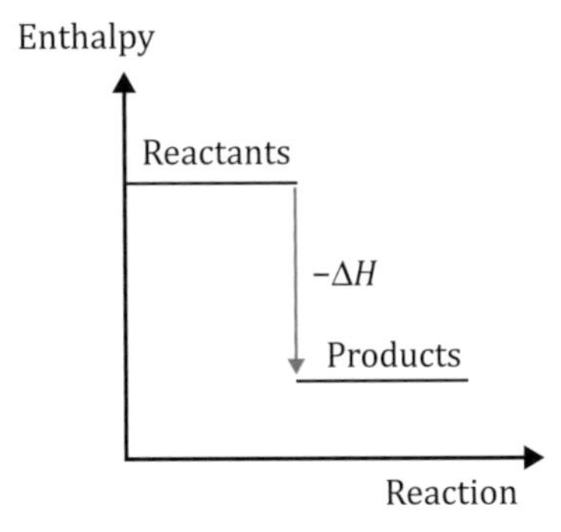

The enthalpy decreases in an exothermic reaction.

Hence ΔH is negative.

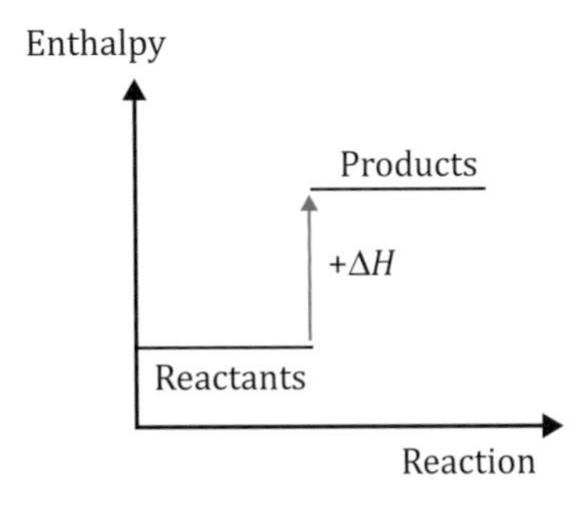

The enthalpy increases in an endothermic reaction.

Hence ΔH is positive.

Examples

1 When a sample of 0.022 mol of methanol was burned completely in air, the heat from the reaction raised the temperature of 140 g of water by 15 °C.

Calculate the enthalpy change in kJ mol^{-1} if one mole of methanol was burned.

(The specific heat capacity of water is 4.18 J K^{-1}g^{-1}.)

Answer

1 As the temperature increases, the reaction is exothermic so the enthalpy change ΔH will be negative.

Heat given to the water, $Q = mc\Delta T$

$$= 140 \times 4.18 \times 15$$
$$= 8778\,\text{J}$$
$$= 8.778\,\text{kJ}$$

Pointer

To convert from J to kJ you divide by 1000.

This is the heat given out when 0.022 mol of methanol burns.

Hence enthalpy change, $\Delta H = -\dfrac{8.778}{0.022} = -399\,\text{kJ mol}^{-1}$

Pointer

To find the enthalpy change in kJ mol^{-1} you divide the energy change in kJ by the number of moles. As the reaction is exothermic, you must remember to add the minus sign as ΔH is always negative for exothermic reactions.

2 When 25.0 g of hydrated copper(II) sulfate ($CuSO_4.5H_2O$) was dissolved in 200 g of water, the temperature fell by 1.4 °C. Calculate the enthalpy change ΔH in kJ mol^{-1}.

(Assume that the specific heat capacity of water is 4.2 J K^{-1}g^{-1}.)

Answer

2 Heat **lost by** the water, $Q = mc\Delta T$

$$= 200 \times 4.2 \times 1.4$$
$$= 1176\,\text{J}$$
$$= 1.176\,\text{kJ}$$

Pointer

Remember to change from J to kJ.

Molar mass of $CuSO_4.5H_2O = 63.5 + 32.1 + 4 \times 16.0 + 5 \times 18$

$$= 249.6\,\text{g mol}^{-1}$$

Number of moles, $n = \dfrac{m}{M} = \dfrac{25.0}{249.6} = 0.1002$ mol

1.176 kJ is the heat **taken in** when 0.1002 mol of hydrated copper sulfate dissolves.

Hence enthalpy change, $\Delta H = \dfrac{1.176}{0.1002} = 11.7365$

$$= +11.7\,\text{kJ mol}^{-1}$$

Pointer

To find the enthalpy change in kJ mol^{-1} you divide the energy change in kJ by the number of moles. As the reaction is endothermic, you must remember to add the + sign.

Using density to determine mass

The following equation connects the density of a liquid with its mass and volume:

$$\text{Density (in g cm}^{-3}\text{)} = \frac{\text{mass in g}}{\text{volume in cm}^3}$$

Using letters rather than words we have $D = \dfrac{m}{V}$

The above equation can be used to calculate the mass if the density and volume of liquid are known.

Example

Water having a density of 1 g cm^{-3} fills a container of volume 200 cm^3. Determine the mass of the water in the container in grams.

Answer

Rearranging $D = \frac{m}{V}$ for m gives $m = DV$

$$m = DV = 1 \times 200 = 200\,g$$

Standard conditions

Enthalpy changes are usually quoted under standard conditions.

Standard conditions are:

- Temperature is 25 °c or 298 K.
- Pressure is 1 atmosphere.
- Solutions have a concentration of 1 mol dm^{-3}.
- Substances are present in their standard states.

When an enthalpy change occurs under standard conditions it is given the symbol $\Delta H^{\ominus}$.

» Pointer

Standard states means the physical states the substances would be in at 25 °C (i.e. 298 K). For example, this would be a gas for ammonia ($NH_3(g)$), a liquid for water ($H_2O(l)$) and a solid for ($Na(s)$).

The enthalpy change of a reaction $\Delta H^{\ominus}$

The standard enthalpy of a reaction $\Delta H^{\ominus}$ is the enthalpy change when a reaction occurs under standard conditions. For example

$$A + B \longrightarrow C + D \qquad \Delta H^{\ominus} = -125\ kJ\ mol^{-1}$$

This means that when one mole of A and B react under standard conditions, the reaction is exothermic (the negative sign shows this) with the release of 125kJ of heat energy.

The enthalpy change of formation of $\Delta H^{\ominus}_f$

The enthalpy change of formation $\Delta H^{\ominus}_f$ is the enthalpy change when 1 mole of a substance is formed from its constituent elements under standard conditions, when all the reactants and products are in their standard states. It is important to note that the standard enthalpy change of formation of any element in its standard state is zero.

The enthalpy change of combustion $\Delta H^{\ominus}_c$

The enthalpy change of combustion $\Delta H^{\ominus}_c$ is the enthalpy change when 1 mole of a substance is completely burned in oxygen under standard conditions, when all the reactants and products are in their standard states.

16.2 Hess's law

Hess's law states that the enthalpy change accompanying a chemical reaction is the same regardless of the route taken from reactants to products.

Working out the standard enthalpy change of a reaction when all the standard enthalpy changes of formation of $\Delta H^\ominus_f$ for the reactants and products are known

In many cases, you cannot measure the enthalpy change for a reaction directly. Instead Hess's law is used with standard enthalpy changes of formation for the reactants and products as the following example shows.

Example

One way of extracting the metal manganese from its ore Mn_2O_3 is by reducing it using carbon monoxide gas at high temperature. The equation for this reaction is shown here.

$$Mn_2O_3(s) + 3CO(g) \longrightarrow 2Mn(s) + 3CO_2(g)$$

The standard enthalpies of formation for the species shown in the equation above are shown in the following table:

Species	$\Delta H_f^\ominus / kJ\,mol^{-1}$
$Mn_2O_3(s)$	−971
$CO(g)$	−111
$Mn(s)$	0
$CO_2(g)$	−394

a) State why the standard enthalpy of formation of Mn(s) is zero.

b) Using data in the above table, calculate the enthalpy change represented by the equation for the extraction of manganese.

Answer

a) Mn(s) is already an element in its standard state. So no energy is required to form it from its elements.

b)

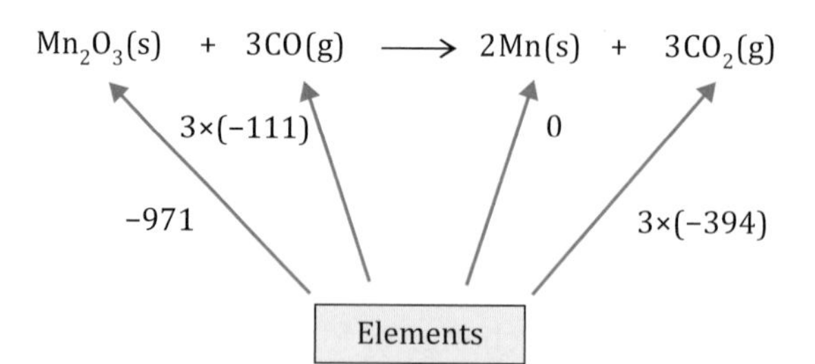

» *Pointer*

Looking at the diagram you can see there is now a path from the reactants on the left, via the elements, to the products and this allows $\Delta H^\ominus$ for the reaction to be found.

Note the way the arrows in the diagram below have been reversed, so that they now go from the reactants to the elements. When this is done the sign of the enthalpy change is reversed.

Hence we now have:

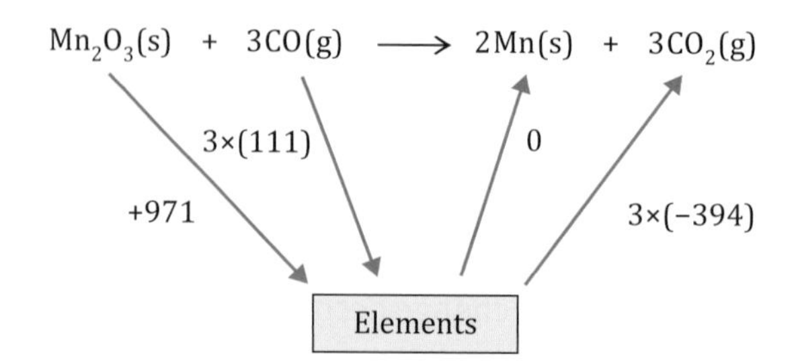

$\Delta H^{\ominus} = 971 + (3 \times +111) + 0 + (3 \times -394)$

$= +122\ \text{kJ mol}^{-1}$

» Pointer

Always include the sign. So we write +122 rather than just 122.

Working out the enthalpy change of a reaction when all the enthalpy changes of combustion $\Delta H^{\ominus}_c$ for the reactants and products are known

The enthalpy change for a reaction can also be found indirectly by using Hess's law and standard enthalpy changes of combustion for the reactants and products as the following example shows.

Example

The standard enthalpy change of formation, $\Delta H^{\ominus}_f$ of glucose, $C_6H_{12}O_6$ cannot be determined directly. Instead it has to be found using standard enthalpy change of combustion data.

The following table shows some enthalpy changes of combustion, ΔH_c.

Substance	$\Delta H^{\ominus}_c$/kJ mol^{-1}
C(s)	−394
H_2(g)	−286
$C_6H_{12}O_6$(s)	−2801

a) Explain what is meant by the standard enthalpy change of formation $\Delta H^{\ominus}_f$.

b) Write down a fully balanced equation for the standard enthalpy change of formation of $C_6H_{12}O_6$.

c) Use the values of ΔH_c in the above table and the equation in your answer to part (a) to calculate the standard enthalpy change of formation of glucose ($C_6H_{12}O_6$).

Answer

a) The standard enthalpy change of formation $\Delta H^{\ominus}_f$ is the enthalpy change when 1 mole of a substance is formed from its constituent elements under standard conditions, when all the reactants and products are in their standard states.

b) $6C(s) + 6H_2(g) + 3O_2(g) \longrightarrow C_6H_{12}O_6(s)$

c)

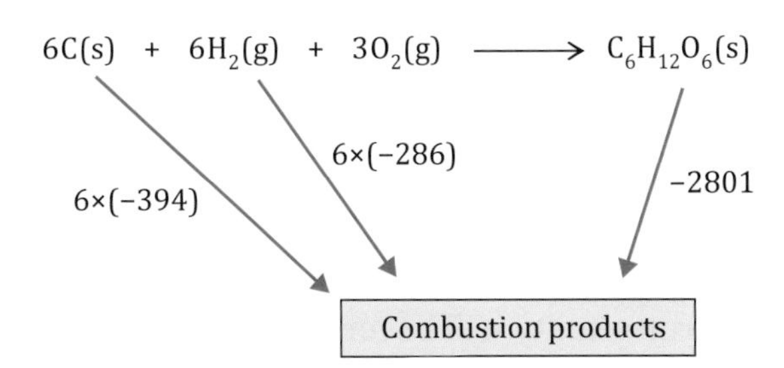

$\Delta H^{\ominus}_f = 6 \times (-394) + 6 \times (-286) + 2801$

$= -1279\ \text{kJ mol}^{-1}$

» Pointer

Remember you need a path from the reactants to the products via the combustion products. The −2801 needs to be changed to +2801 as the arrow needs to go in the opposite direction.

16.3 Born–Haber cycles

Some enthalpy changes, such as the lattice formation enthalpy, cannot be measured directly by experiment. It is possible to use known enthalpy changes, obtained by experiment or looked up in tables, to calculate an unknown enthalpy change using what is called a Born–Haber cycle.

In order to construct Born–Haber cycles it is necessary to understand the following definitions.

» Pointer
Standard conditions are a pressure of 100 kPa and usually a temperature of 298 K.

The standard enthalpy change of formation $\Delta H^\ominus_f$ is the enthalpy change when one mole of a substance is formed from its constituent elements under standard conditions, when all the reactants and products are in their standard states. It is important to note that the standard enthalpy change of formation of any element in its standard state is zero.

The standard enthalpy change of atomisation $\Delta H^\ominus_{at}$ is the enthalpy change under standard conditions, when one mole of gaseous atoms is formed from the element in its standard state.

» Pointer
Notice that the one mole in these reactions refers to the element on the right of the equation.

Here are some examples involving the enthalpy change of atomisation:

$$\tfrac{1}{2}Cl_2(g) \longrightarrow Cl(g)$$

$$Na(s) \longrightarrow Na(g)$$

The first ionisation energy is the enthalpy change under standard conditions, when one mole of gaseous atoms is converted into one mole of singly positively charged gaseous ions.

Here are some examples where the first ionisation energy is involved:

Pointer
For Group 2 elements such as Mg and Ca, there will be two ionisation energies involved. The first ionisation energy is associated with removal of the first electron and the second ionisation energy is associated with the removal of the second electron to produce the M^{2+} ion.

$$Na(g) \longrightarrow Na^+(g) + e^-$$

$$Ca(g) \longrightarrow Ca^+(g) + e^-$$

The second ionisation energy is the enthalpy change under standard conditions, when one mole of gaseous singly charged positive ions is converted into one mole of gaseous doubly charged positive ions.

Here are some examples where the second ionisation energy is involved:

$$Ca^+(g) \longrightarrow Ca^{2+}(g) + e^-$$

$$Mg^+(g) \longrightarrow Mg^{2+}(g) + e^-$$

The electron affinity, $\Delta H^\ominus_{ea}$ is the enthalpy change under standard conditions when one mole of gaseous atoms is converted to one mole of singly negatively charged gaseous ions.

Note that electron affinity is sometimes referred to electron attachment enthalpy.

If the ion being formed has a −2 charge then there will be two electron affinities involved. The first electron affinity will be for the reaction:

$X(g) + e^- \longrightarrow X^-(g)$ Enthalpy change is the first electron affinity

and the second electron affinity will be for the reaction:

$X^-(g) + e^- \longrightarrow X^{2-}(g)$ Enthalpy change is the second electron affinity

Note that the enthalpy changes for the first and second electron affinities are different.

The standard lattice formation enthalpy $\Delta H^{\ominus}_{L}$ is the enthalpy change under standard conditions when one mole of a solid ionic compound is formed from its constituent ions in their gaseous state. The lattice formation enthalpy is always exothermic so the enthalpy change is negative.

The standard lattice dissociation enthalpy $\Delta H^{\ominus}_{L}$ is the reverse of the standard lattice formation enthalpy, and is the enthalpy change under standard conditions when one mole of a solid ionic compound is changed to ions in their gaseous state. Numerically it has the same value as the standard lattice formation enthalpy but has an opposite sign.

Using a Born–Haber cycle to determine the lattice formation enthalpy $\Delta H^{\ominus}_{L}$ for sodium chloride

Suppose we want to find the lattice formation enthalpy for sodium chloride. It is necessary for the solid sodium chloride to be turned to gaseous ions and there are a number of steps taken to achieve this. We need to first follow a series of steps where the enthalpy changes are known and then draw these as a circular diagram called a Born–Haber cycle. We then use Hess's law to determine the lattice formation enthalpy or a different unknown enthalpy change.

Pointer
You will find that many of the standard enthalpy values vary between exam questions. You must always use the values given in the question or looked up in data booklets given.

We need to think about the state and arrangement of atoms of sodium and chlorine under standard conditions. Sodium will be solid and consist of atoms and chlorine will be a gas consisting of chlorine molecules each having two atoms of chlorine.

This is the starting point for the Born–Haber cycle and it can be represented on the diagram in the following way:

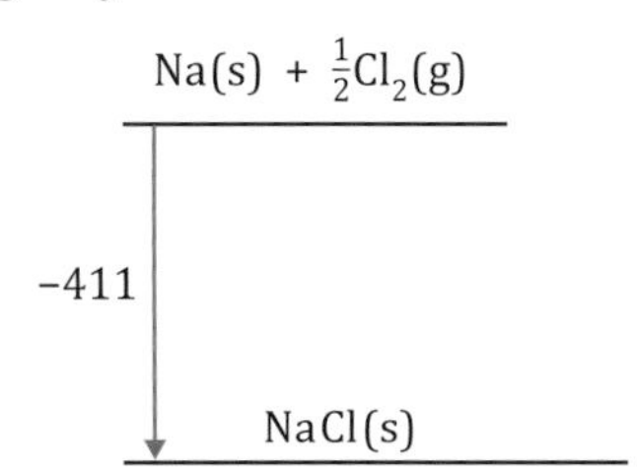

Pointer
The standard enthalpy of formation of sodium chloride is $-411\,kJ\,mol^{-1}$.

The standard enthalpy of formation of NaCl is represented by the following equation and the enthalpy change will be given or can be looked up in tables:

$$Na(s) + \tfrac{1}{2}Cl_2(g) \longrightarrow NaCl(s) \qquad \Delta H^{\ominus}_{f} = -411\ kJ\ mol^{-1}$$

Pointer
NaCl(s) is placed on a long horizontal line as this is the substance being produced. Notice that a negative enthalpy change is given an arrow going down. Positive enthalpy changes will go up.

We now need to complete a cycle starting from solid sodium and gaseous chlorine molecules and eventually getting back to solid sodium chloride on the bottom line of the above diagram.

The first step will be to convert both sodium and chlorine to gaseous atoms. This will involve the standard enthalpy change of atomisation and these will normally be given or can be looked up in tables.

The equations are as follows:

$$Na(s) \longrightarrow Na(g) \qquad \Delta H^{\ominus}_{at} = +108\ kJ\ mol^{-1}$$

$$\tfrac{1}{2}Cl_2(g) \longrightarrow Cl(g) \qquad \Delta H^{\ominus}_{at} = +122\ kJ\ mol^{-1}$$

Pointer
Note that here we are forming one mole of gaseous atoms.

These enthalpy changes are represented on the diagram by lines with upward arrows:

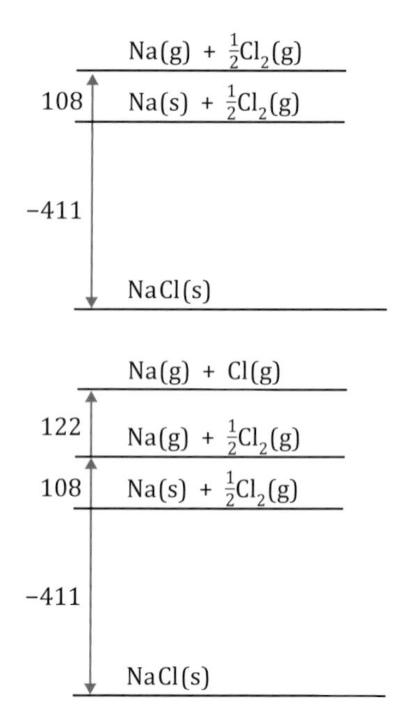

Now both elements consist only of gaseous atoms. The next stage is to ionise the sodium to Na^+ ions. This will involve the first ionisation energy, $\Delta H^{\ominus}_{i}$ and this will normally be given or it can be looked up in tables.

The equation for the ionisation of sodium is as follows:

$$Na(g) \longrightarrow Na^+(g) + e^- \qquad \Delta H^{\ominus}_{i} = +494\ \mathrm{kJ\ mol^{-1}}$$

This is added to the diagram:

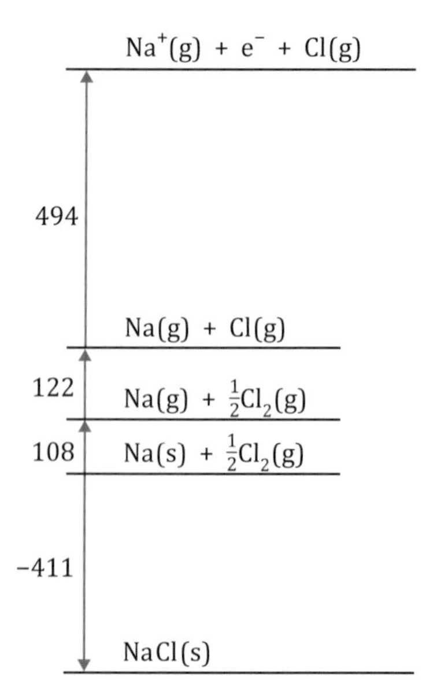

Next the chlorine is ionised by accepting an electron and this involves the electron affinity, $\Delta H^{\ominus}_{ea}$.

The equation for the electron affinity for chlorine is as follows:

$$Cl(g) + e^- \longrightarrow Cl^-(g) \qquad \Delta H^{\ominus}_{i} = -364\ \mathrm{kJ\ mol^{-1}}$$

We now add this with the arrow pointing downwards to the diagram as follows:

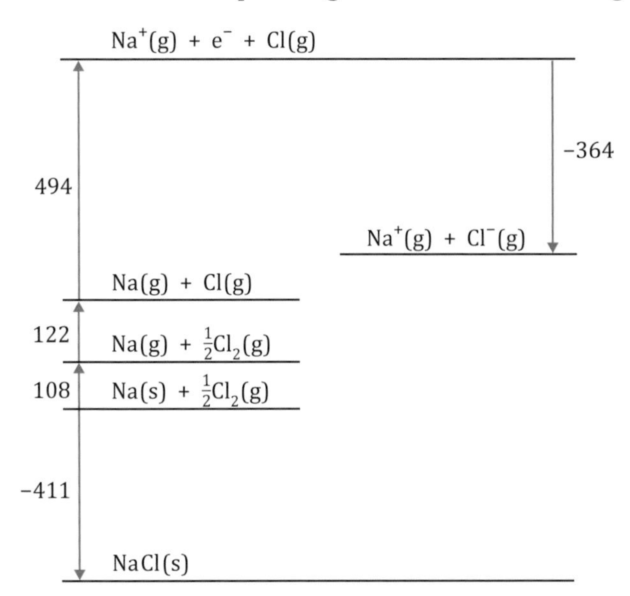

Now both the sodium and chlorine consist of gaseous ions we need to combine to form solid sodium chloride. This will involve the standard lattice formation enthalpy, $\Delta H^{\ominus}_{L}$.

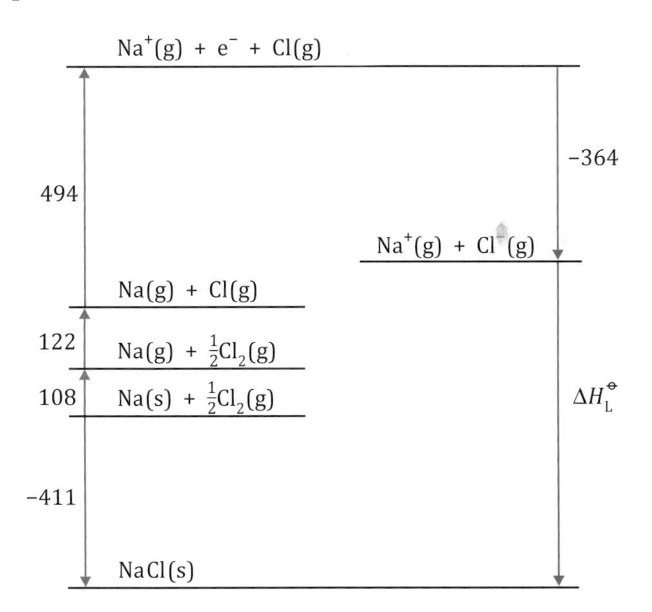

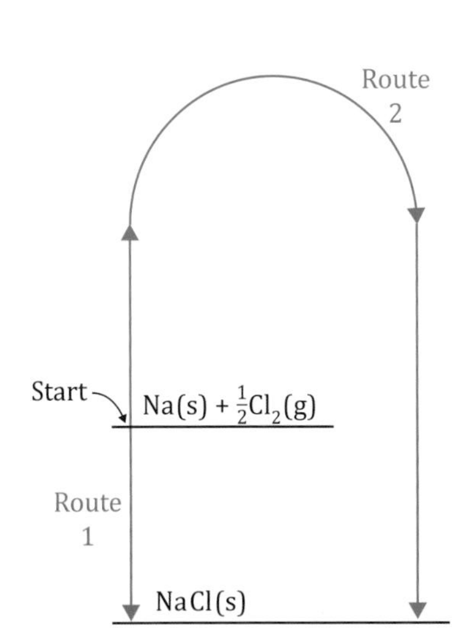

We now completed the Born–Haber cycle. If you look at the simplified diagram right, you will see there are two routes by which NaCl(s) can be created. The simple route 1 involves the standard enthalpy of formation whilst route 2 involves many steps but according to Hess's law the enthalpy change will be the same as for route 1.

Hence we can create the following equation.

Route 1 = Route 2

So, $-411 = 108 + 122 + 494 + (-364) + \Delta H^{\ominus}_{L}$

Solving, gives $\Delta H^{\ominus}_{L} = -771 \text{ kJ mol}^{-1}$

» Pointer

Although we have not added the units to the numbers on the Born–Haber cycle, we must remember to include them in the final answer. The units are in kJ mol^{-1}.

Example

The Born–Haber cycle shown below shows the standard enthalpy changes involved in the formation of magnesium oxide (MgO).

The diagram is not drawn to scale and all the standard enthalpy changes are in the units kJ mol^{-1}.

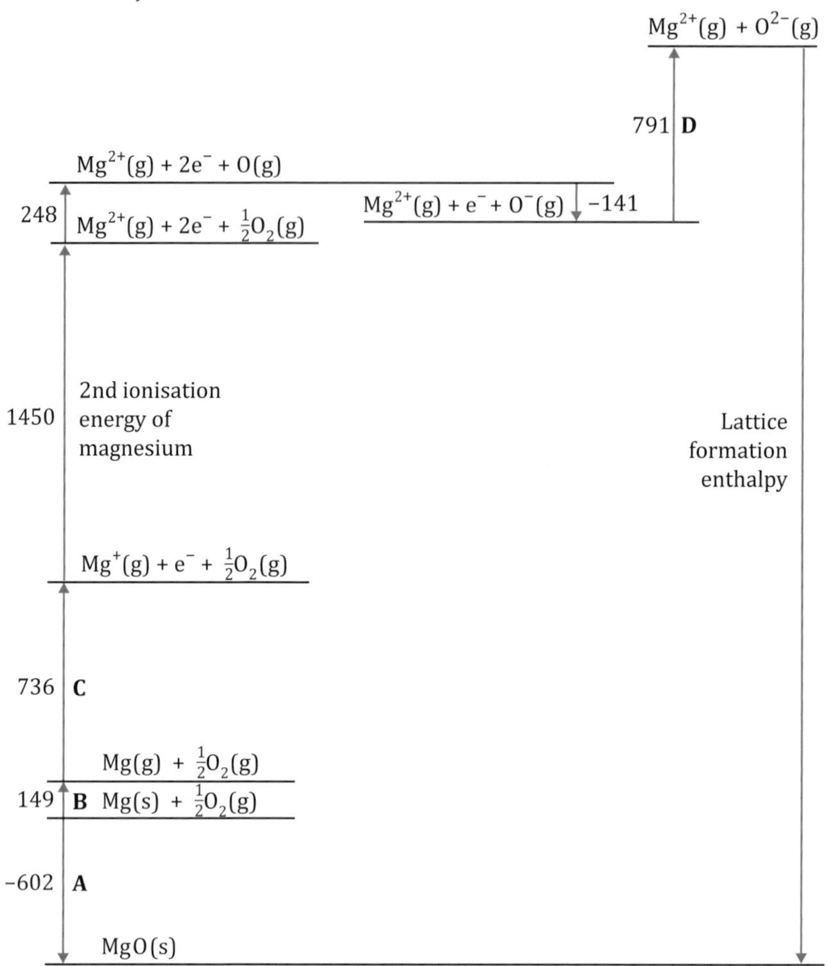

a) Give the names of the standard enthalpy changes marked with the following letters.

i) A

ii) B

iii) C

iv) D

b) Using the data from the Born–Haber cycle, calculate the value of the standard lattice formation enthalpy.

Answer

a) i) A is the standard enthalpy of formation of magnesium oxide.

ii) B is the standard enthalpy of atomisation of magnesium.

iii) C is the first ionisation energy for magnesium.

iv) D is the second electron affinity for oxygen.

b) $-602 = 149 + 736 + 1450 + 248 + (-141) + 791 + \Delta H^{\ominus}_{L}$

Solving gives = −3835 kJ mol^{-1}

Pointer

We equate the energy changes for the two routes to produce magnesium oxide.

Enthalpy change of solution

When an ionic compound dissolves in water there is an enthalpy change which is called the enthalpy change of solution.

The standard enthalpy change of solution ΔH°_{sol} is the enthalpy change under standard conditions when one mole of a solid compound dissolves completely in water.

When an ionic solid dissolves completely in water the following happens.

- The ionic lattice breaks forming separate gaseous ions. To do this requires energy and the enthalpy change for this is called the standard lattice dissociation enthalpy, ΔH°_{lat}. Note that the standard lattice dissociation energy has the same size as the standard lattice formation enthalpy but opposite sign. The standard lattice dissociation enthalpy change is endothermic and the standard lattice formation enthalpy change is exothermic.
- The positive ions are hydrated. The energy released is called the standard enthalpy of hydration, ΔH°_{hyd} of the positive ions.
- The negative ions are hydrated. The energy released is called the standard enthalpy of hydration, ΔH°_{hyd} of the negative ions.

For sodium chloride, ΔH°_{sol} would be the overall standard enthalpy change for the following:

$$NaCl(s) \longrightarrow Na^+(g) + Cl^-(g) \longrightarrow Na^+(aq) + Cl^-(aq)$$

Pointer

Notice that there are two ions to hydrate so there will be two standard enthalpies of hydration.

Hence we can write $\mathbf{\Delta H^\circ_{sol} = \Delta H^\circ_{lat} + \Delta H^\circ_{hyd}}$ **of** $\mathbf{Na^+}$ + $\mathbf{\Delta H^\circ_{hyd}}$ **of** $\mathbf{Cl^-}$

In the case of an ionic compound such as $CaCl_2$, when working out the enthalpy of hydration, you would need to multiply the standard enthalpy of hydration of the chloride ion by 2 as two chloride ions would be hydrated.

Example

If the standard enthalpy of hydration of $F^-(g)$ ions is −506 kJ mol^{-1}, the standard enthalpy of hydration of $Ag^+(g)$ ions is −464 kJ mol^{-1} and the standard enthalpy of solution for silver fluoride (AgF) in water is −20 kJ mol^{-1}, calculate the standard lattice enthalpy of dissociation of silver fluoride.

Answer

Standard enthalpy of solution = standard lattice enthalpy of dissociation of silver fluoride + standard enthalpy of hydration of $Ag^+(g)$ + standard enthalpy of hydration of $F^-(g)$

Hence, substituting the values in, we have:

−20 = standard lattice enthalpy of dissociation of silver fluoride + (−464) + (−506)

−20 = standard lattice enthalpy of dissociation of silver fluoride − 970

Hence, standard lattice enthalpy of dissociation of silver fluoride = 970 − 20 = 950 kJ mol^{-1}

Test yourself 16

1. When 12.0 g anhydrous copper(II) sulfate ($CuSO_4$) was dissolved in 100 g water a temperature rise of 1.8 °C was recorded. Calculate the standard enthalpy change $\Delta H^\ominus$ in kJ mol^{-1}.

   (The specific heat capacity of water is 4.2 J g^{-1} K^{-1})

2. Here is a table of data relating to certain reactions:

Reaction	Standard enthalpy change/kJ mol^{-1}
$H(g) + Cl(g) \longrightarrow HCl(g)$	−432
$HCl(g) \longrightarrow H^+(aq) + Cl^-(aq)$	−75
$H(g) + Cl(g) \longrightarrow H^+(g) + Cl^-(g)$	+963

   Here is a scheme for a reaction:

$$\begin{array}{ccc} H^+(g) + Cl^-(g) & \xrightarrow{\Delta H_r^\ominus} & H^+(aq) + Cl^-(aq) \\ \uparrow & & \uparrow \\ H(g) + Cl(g) & \longrightarrow & HCl(g) \end{array}$$

   Use the scheme of reactions and the data in the table to calculate a values for $\Delta H^\ominus{}_r$.

3. Methanol burns in air according to the following reaction:

   $CH_3OH(l) + 1\frac{1}{2}O_2(g) \longrightarrow CO_2(g) + 2H_2O(l)$

   The following table shows the standard enthalpy change of formation for some the reactants and products in the above reaction.

Compound	Standard enthalpy change of formation, $\Delta H_f^\ominus$/kJ mol^{-1}
$CH_3OH(l)$	−239
$CO_2(g)$	−394
$H_2O(l)$	−286

   (a) State the standard enthalpy change of formation for $O_2(g)$.

   (b) Calculate the standard enthalpy change of combustion for methanol.

4. The standard lattice dissociation enthalpy of magnesium chloride ($MgCl_2$) is 2493 kJ mol^{-1}. When magnesium chloride dissolves in water, the standard enthalpy of hydration of the magnesium ions is −1920 kJ mol^{-1}and the standard enthalpy of solution is −155 kJ mol^{-1}.

   Calculate the standard enthalpy of hydration of the chloride ions.

5. The Born–Haber cycle shown on page 103 shows the standard enthalpy changes involved in the formation of sodium iodide (NaI).

   The diagram is not drawn to scale and all the standard enthalpy changes are in the units kJ mol^{-1}.

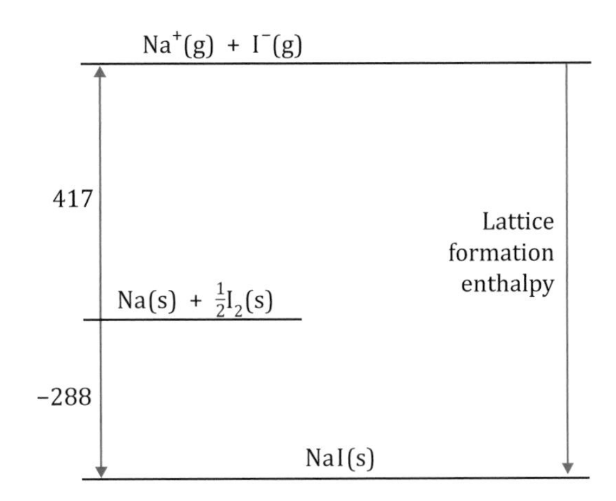

Using data in the diagram, calculate the lattice formation enthalpy, $\Delta H^\ominus_L$ for sodium iodide.

6 The Born–Haber cycle shown below shows the standard enthalpy changes involved in the formation of magnesium chloride ($MgCl_2$).

The diagram is not drawn to scale and the standard enthalpy changes are in the units kJ mol^{-1}.

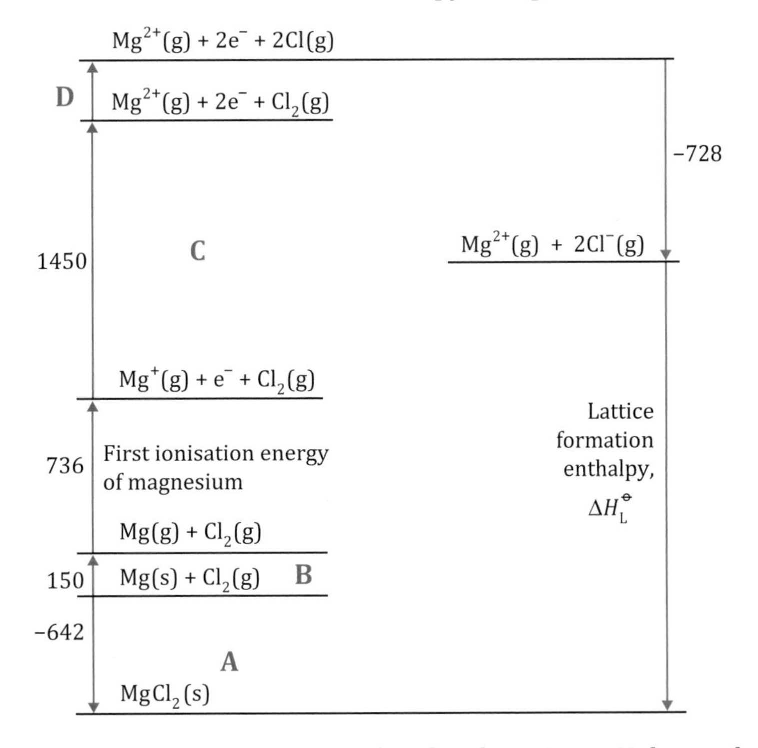

(a) Here are some of the energy changes represented in the above Born–Haber cycle:

Second ionisation energy of magnesium

Standard enthalpy change of formation of magnesium chloride

Standard enthalpy change of atomisation of magnesium

From the list above and by reference to the diagram, state the name of the energy change written by each of the following letters on the diagram:

(i) A

(ii) B

(iii) C

(b) The standard atomisation energy of chlorine has the equation and enthalpy change as follows:

$\frac{1}{2}Cl_2(g) \longrightarrow Cl(g) \qquad \Delta H^{\circ}_{at} = +121\ kJ\ mol^{-1}$

Calculate the energy change represented by the letter D on the diagram.

(c) Use the data on the diagram and your answer to part (b) to calculate the standard lattice formation enthalpy for magnesium chloride.

Chapter 17 Bond enthalpies

Energy is needed to break bonds, so bond breaking is endothermic. Energy is released when bonds are formed so bond forming is exothermic. In a particular reaction the reactants have their bonds broken and these re-form to give the products. The difference between the energies involved in bond breaking and bond forming will determine whether a particular reaction is exothermic or endothermic.

Hence we have the following equation for enthalpy change:

$$\text{Enthalpy change for reaction} = \text{Enthalpy change in breaking bonds} - \text{Enthalpy change in forming bonds}$$

» *Pointer*
Make sure you remember this formula.

The bond enthalpy is the energy required to break or form a covalent bond with both the species involved being in the gaseous state. The units for bond enthalpy are kJ mol^{-1}.

Bond enthalpies can be used to work out an enthalpy change for a reaction as the following example shows.

Example

Use the bond enthalpies in the following table to work out the enthalpy change for the following reaction:

$$2H_2(g) + O_2(g) \longrightarrow 2H_2O(g)$$

Bond	Bond enthalpy/ kJ mol^{-1}
H—H	436
O=O	497
O—H	463

Answer

First write the equation showing all the covalent bonds:

H—H + O=O ⟶ H—O—H
H—H   H—O—H

Enthalpy change when bonds are broken:

$2 \times$ H—H $= 2 \times 436 = 872$ kJ mol^{-1}

$1 \times$ O=O $= 497$ kJ mol^{-1}

Total enthalpy change in breaking bonds $= 872 + 497 = 1369$ kJ mol^{-1}

Enthalpy change when bonds are formed:

$4 \times$ O—H $= 4 \times 463 = 1852$ kJ mol^{-1}

$$\text{Enthalpy change for reaction} = \text{Enthalpy change in breaking bonds} - \text{Enthalpy change in forming bonds}$$

$= 1369 - 1852$

$= -483$ kJ mol^{-1}

17.1 Using the enthalpy change for a reaction to work out a bond enthalpy

If the enthalpy change is known for a reaction and some of the bond enthalpies are known, then an unknown bond enthalpy can be found, as this example illustrates.

Example

The addition of hydrogen to ethene is an exothermic reaction whose equation is shown below:

```
                             H  H
H       H                    |  |
 \     /                     |  |
  C═C       + H—H  ───→   H—C—C—H
 /     \                     |  |
H       H                    |  |
                             H  H      ΔH = –136 kJ mol⁻¹
```

The following table shows some bond enthalpy data for some of the bonds involved in the above reaction:

Bond	Bond enthalpy/kJ mol^{-1}
C═C	612
C—C	348
H—H	436

Use the information from the equation and the table of data to calculate the bond enthalpy for the C–H bond.

Answer

Enthalpy change when bonds are broken:

= 4 × C—H + 1 × C═C + 1 × H—H

= (4 × C—H) + (1 × 612) + (1 × 436)

= 4 × C—H + 1048

Enthalpy change when bonds are formed:

= 6 × C—H + 1 × C—C

= (6 × C—H) + (1 × 348)

= 6 × C—H + 348

Enthalpy change for reaction = Enthalpy change in breaking bonds – Enthalpy change in forming bonds

–136 = (4 × C—H + 1048) – (6 × C—H + 348)

–136 = –2C—H + 700

2C—H = 700 + 136

2C—H = 836

C—H = +418 kJ mol^{-1}

» *Pointer*
Bond breaking is always endothermic so as a check this value will be positive. You should always add the + sign.

Test yourself 17

1. Using the bond enthalpies in the following table, calculate the energy released when methane is burnt in a plentiful supply of oxygen. The equation for the reaction is:

$$CH_4(g) + 2O_2(g) \longrightarrow CO_2(g) + 2H_2O(g)$$

Bond	Bond enthalpy/kJ mol^{-1}
C—H	413
O=O	497
O—H	463
C=O	805

2. Use the data in the table below to work out the enthalpy change for the reaction of ethane with chlorine according to the following reaction:

$$C_2H_6(g) + Cl_2(g) \longrightarrow C_2H_5Cl(g) + HCl(g)$$

Bond	Bond enthalpy/kJ mol^{-1}
C—H	413
C—C	347
C—Cl	346
Cl—Cl	243
H—Cl	432

3. Here is a table containing mean bond enthalpy data:

Bond	Bond enthalpy/kJ mol^{-1}
N≡N	+944
H—H	+436
N—H	+388

(a) Write an equation to create one mole of ammonia, NH_3, from its elements in their standard states.

(b) Using data in the above table, calculate a value for the enthalpy of formation of ammonia.

4. Hydrazine, N_2H_4, was mixed with hydrogen peroxide, H_2O_2, to make rocket fuel.
The reaction between hydrazine and hydrogen peroxide is shown here:

$$H_2N{-}NH_2 + 2\,H{-}O{-}O{-}H \longrightarrow N{\equiv}N + 4\,H{-}O{-}H$$

Using the data in the table below, calculate the enthalpy change for the reaction as shown above. You should assume that both reactants and products are gaseous.

Bond	Bond enthalpy/kJ mol^{-1}
N—H	388
N—N	163
N≡N	944
O—H	463
O—O	146

Chapter 18 Rates of reaction

18.1 Definition for rate of reaction

The rate of reaction is defined as the change in concentration of any of the reactants or products per unit time. Rate of reaction is usually measured in the units mol dm^{-3} s^{-1}.

Suppose we have the following reaction:

$$A + 2B \longrightarrow C$$

The rate of the above reaction is given by:

$$\text{rate} \propto [A]^m[B]^n$$

Removing the proportion sign and including a constant of proportionality, called the rate constant, *k*, we have

$$\text{rate} = k[A]^m[B]^n$$

It is important to note that the rate constant, *k*, varies with temperature, so the temperature at which the reaction takes place remains constant.

The indices (powers) *m* and *n* are called the orders of the reaction with respect to the reactants A and B respectively.

The overall order of a reaction is the sum of the indices (i.e. powers) of the concentrations in the rate equation. For the above equation, the overall order = $m + n$.

» Pointer
The symbol ∝ means proportional to.

» Pointer
Note that *m* and *n* do not refer to the stoichiometry of the equation and must be determined by experiment.

» Pointer
[A] means the concentration of A.

» Pointer
This is referred to as the rate equation.

18.2 Examples in the use of the rate equation

Suppose a reaction had the following rate equation:

$$\text{rate} = k[A][B]$$

If the concentration of A doubled and B stayed the same, the rate would double. If the concentrations of A and B doubled, the rate would quadruple.

Suppose another reaction had the following rate equation:

$$\text{rate} = k[A]^2[B]$$

If the concentration of A doubled and B stayed the same, the rate would quadruple, as the concentration of A is squared in the rate equation. If the concentration of A halved and B stayed the same, the rate would be quartered.

Suppose [A] is doubled and [B] is halved. Doubling [A] would quadruple the rate and halving [B] would halve the rate so applying both of these would give a net doubling of the rate.

Zero-order reactions

Suppose we had a reaction which had the rate equation:

$$\text{rate} = k[\text{A}]^x$$

If the rate did not depend on the concentration of A, then $x = 0$ and the reaction is a zero-order reaction.

So, $$\text{rate} = k[\text{A}]^0 = k$$

> **Pointer**
> Any number (or letter representing a number) raised to the power of zero is equal to one.
> Hence $[\text{A}]^0 = 1$.

This means that the rate of this reaction does not depend on the concentration of A.

k will have the same units as the rate (i.e. mol $\text{dm}^{-3}\text{s}^{-1}$).

The graph of [A] against time for a zero-order reaction is shown below:

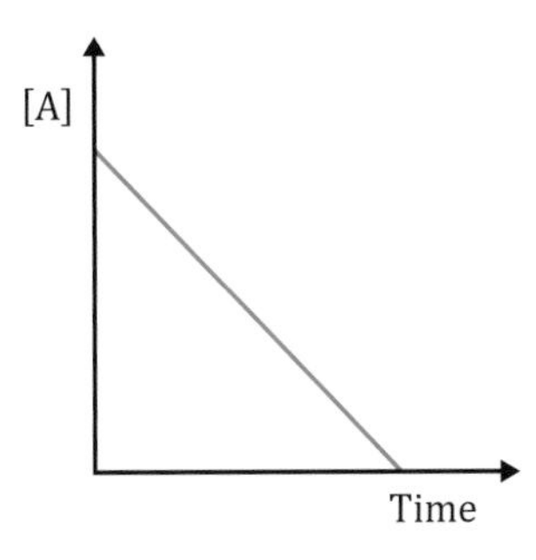

> **Pointer**
> The gradient of this graph represents the rate of reaction. As the gradient is constant, the rate will be constant. This can be seen in the second graph where the rate is plotted against time.

If a graph of rate against time is plotted, it results in a line parallel to the x-axis showing the rate stays constant. This graph is shown below:

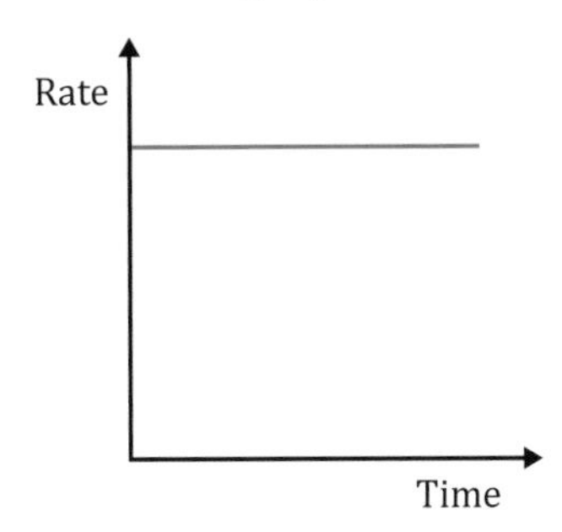

First-order reactions

In a first-order reaction the rate equation could be:

$$\text{rate} = k[\text{A}]^1 = k[\text{A}]$$

The rate is proportional to the concentration of A, so if [A] doubles then the rate will double, if [A] halves, then the rate will halve and so on.

The graph of [A] against time for a first-order reaction is shown on page 110:

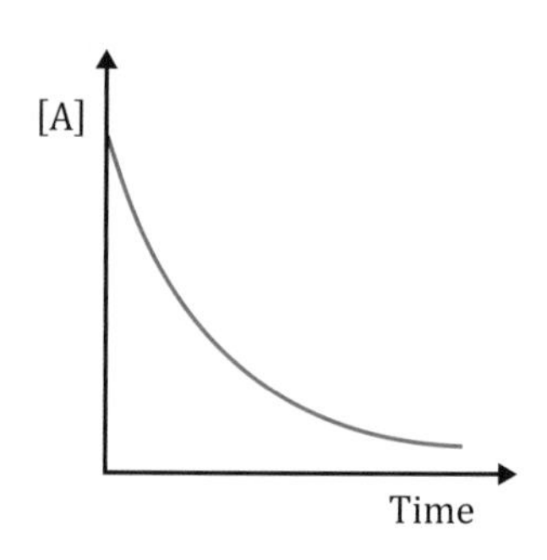

If a graph of rate against concentration is plotted, it results in a straight line passing through the origin showing that the rate of reaction is directly proportional to the concentration.

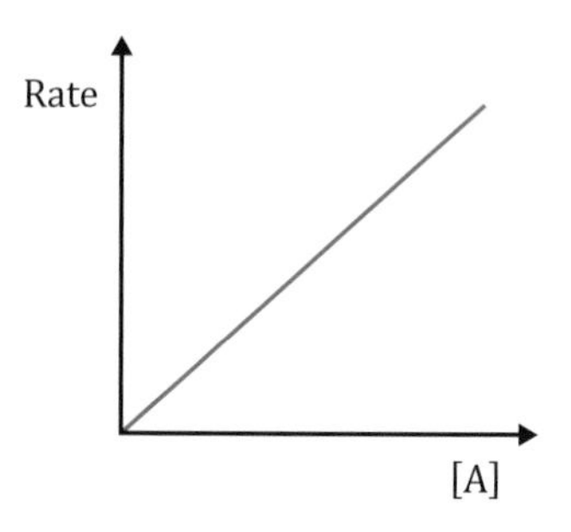

»Pointer
The graph of Rate against [A] is a straight line through the origin. This means that the rate is directly proportional to the concentration.

Second-order reactions

In a second-order reaction the rate equation could be:

$$\text{rate} = k[\text{A}]^2$$

The rate is proportional to the square of the concentration of A. If [A] doubles then the rate will quadruple, if [A] halves, then the rate will be one quarter of its original value and so on.

»Pointer
Note that the order of the reaction is 2 with respect to A. The overall rate of reaction is also 2.

The graph of [A] against time for a second-order reaction is shown below:

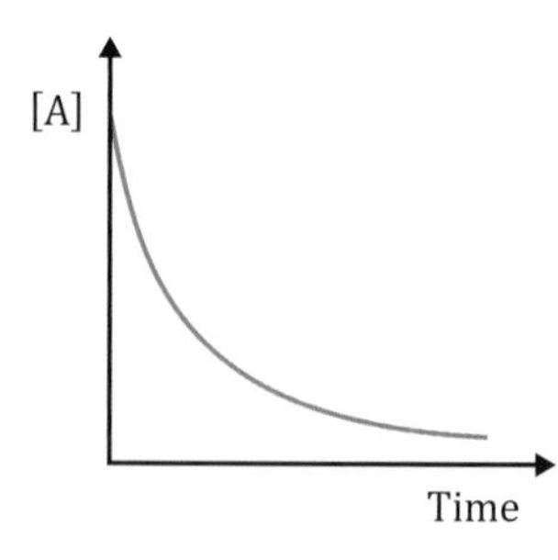

The graph of rate against concentration for a second-order reaction is a curve as shown here:

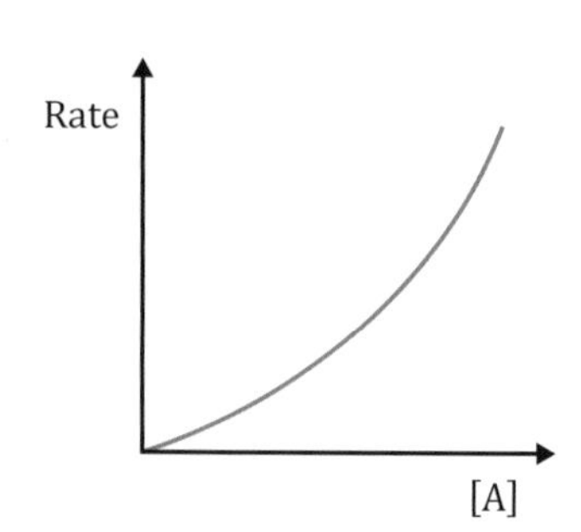

Suppose we had a reaction where the rate was directly proportional to the initial concentration of A and the initial concentration of B and no other concentrations affected the rate, the rate equation would be:

$$\text{rate} = k[A]^1[B]^1$$

> **Pointer**
> The powers of one have been included here to show that the reaction is first order in respect to [A] and [B]. These powers could be left out.

Hence the rate is first order with respect to the initial concentration of A (i.e. [A]) and also to the initial concentration of B (i.e. [B]). The overall order would be 1 + 1 = 2, so this is a second-order reaction.

18.3 Finding a rate of reaction using a concentration–time graph

To find the rate of reaction at a particular time, the gradient of the concentration–time graph is found by drawing a tangent to the curve at the time and then drawing a triangle to determine the gradient.

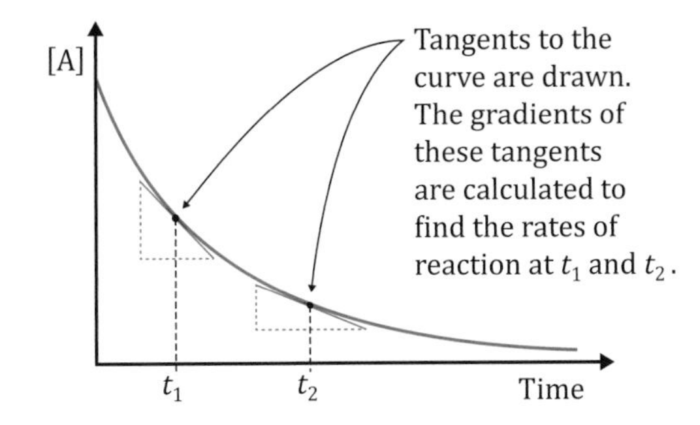

18.4 Determining the units of the rate constant, *k*

The units for the rate constant, *k*, vary depending on the rate equation. The units for a concentration before it is raised to any powers will be mol dm^{-3} and the rate will always have the units mol dm^{-3}s^{-1}. The units for the rate constant are found by rearranging the rate equation to make *k* the subject and then substituting the units into the equation and cancelling units where possible.

For example, suppose a reaction had a rate equation given by:

$$\text{rate} = k[A][B]^2$$

> **Pointer**
> To rearrange this equation so that k becomes the subject, we divide both sides of the equation by $[A][B]^2$

Changing the subject of the equation to *k*, we obtain

$$k = \frac{\text{rate}}{[A][B]^2}$$

Now we put the units in for the rate and the concentrations:

$$k = \frac{\text{mol dm}^{-3}\text{s}^{-1}}{[\text{mol dm}^{-3}]\,[\text{mol dm}^{-3}]^2} = \frac{\text{s}^{-1}}{\text{mol}^2\,\text{dm}^{-6}} = \text{mol}^{-2}\,\text{dm}^6\,\text{s}^{-1}$$

> **Pointer**
> When a unit on the bottom of a fraction is brought to the top, the sign of the power changes so, for example, mol^2 on the bottom would become mol^{-2} on the top.

Example

Two experiments were conducted concerning the reaction between two substances A and B at constant temperature and the following data were obtained:

Experiment	Initial concentration of A /mol dm^{-3}	Initial concentration of B/mol dm^{-3}	Initial rate/ mol $dm^{-3}s^{-1}$
1	2.2×10^{-2}	1.2×10^{-2}	3.6×10^{-5}
2	3.3×10^{-2}	3.6×10^{-2}	To be calculated

The rate equation for the reaction between A and B is known to be:

$$\text{rate} = k[A]^2[B]$$

a) Using the data for experiment 1 in the table, calculate the value of the rate constant, k, at this temperature and deduce its units.

b) Using the rate constant, k, worked out in part (a), calculate the initial rate of reaction for experiment 2.

Answer

a) rate = $k[A]^2[B]$

Rearranging this equation to make k the subject gives:

$$k = \frac{\text{rate}}{[A]^2[B]} = \frac{3.6 \times 10^{-5}}{[2.2 \times 10^{-2}]^2[1.2 \times 10^{-2}]} = 6.20 \text{ (3 s.f.)}$$

To determine the units for k, we put the units for the rate and concentrations into the equation.

$$\text{Hence, } k = \frac{\text{rate}}{[A]^2\,[B]} = \frac{\text{mol dm}^{-3}\text{s}^{-1}}{[\text{mol dm}^{-3}]^2\,[\text{mol dm}^{-3}]} = \frac{\text{s}^{-1}}{[\text{mol dm}^{-3}]^2} = \text{mol}^{-2}\,\text{dm}^{6}\,\text{s}^{-1}$$

b) rate = $k[A]^2[B] = 6.20 \times (3.3 \times 10^{-2})^2 \times 3.6 \times 10^{-2} = 2.43 \times 10^{-4}$ mol $dm^{-3}s^{-1}$

18.5 Determining the rate equation using the initial rate method

When a reaction proceeds, reactants get used up so their concentration decreases with time. It is hard to measure the concentration at a particular time accurately. However, it is easy to measure the concentration of the reactants at the start of the reaction, so the initial rate of reaction is easy to determine. By varying the concentrations and seeing how the initial rate of reaction changes as a result, the rate equation for a reaction can be found as the following example shows.

Example

Three experiments were performed between substances B and C to investigate the initial rates of reaction. All the experiments were conducted at the same constant temperature, and the data for each experiment is shown in the following table.

Experiment	Initial concentration of B/mol dm^{-3}	Initial concentration of C/mol dm^{-3}	Initial rate/ mol dm^{-3}s^{-1}
1	0.15	0.24	0.11×10^{-3}
2	0.45	0.24	0.99×10^{-3}
3	0.90	0.48	3.96×10^{-3}

a) Using the data in the above table, deduce the order of reaction with respect to B.

b) Using the data in the above table, deduce the order of the reaction with respect to C.

Answer

a) Comparing experiments 1 and 2 you can see that the initial concentration of B increases from 0.15 to 0.45. This is an increase by a factor of 3. Notice that the initial concentration of C stays constant at 0.24 and will thus not affect the rate.

The increase in the initial concentration of B by a factor 3 does not increase the initial rate by a factor of 3. Instead the increase is $\frac{0.99 \times 10^{-3}}{0.11 \times 10^{-3}} = 9$.

As 9 is 3^2, this means that the rate depends on the square of the concentration of B.

Hence the reaction is second order with respect to B.

b) Comparing experiments 2 and 3 you can see that the concentrations of both B and C double. The doubling of the concentration of B would quadruple the rate (this has been established in part (a)).

Now the factor by which the rate has increased is $\frac{3.96 \times 10^{-3}}{0.99 \times 10^{-3}} = 4$.

This increase would have been produced by the increase in concentration of B alone which means that the concentration of C has no effect on the rate.

Hence the reaction is zero order with respect to C.

18.6 Finding the rate-determining step for a reaction

Lots of reactions take place in more than one step. In a two-step reaction one of the steps may take place slowly compared to the other. This slow step will influence the rate more than the faster step and is therefore called the rate-determining step.

Example

The equation below shows the reaction between persulfate and iodide ions in aqueous solution

$$S_2O_8^{2-} + 2I^- \longrightarrow 2SO_4^{2-} + I_2$$

The rate equation for the above reaction is:

$$\text{rate} = k[S_2O_8^{2-}][I^-]$$

It has been suggested that the above reaction takes place in two stages:

Step 1 $S_2O_8^{2-} + I^- + H_2O \longrightarrow 2SO_4^{2-} + HOI + H^+$

Step 2 $HOI + H^+ + I^- \longrightarrow H_2O + I_2$

State, giving a reason, why Step 1 is the rate-determining step.

Answer

Look at the rate equation and you can see that the rate depends on the concentrations of the persulfate ions and the iodide ions. Both of these ions appear as reactants in Step 1, so this is the rate-determining step.

18.7 The Arrhenius equation

The rate constant, k, is only constant at a particular temperature. The value of the rate constant, k, at different temperatures can be found using the Arrhenius equation shown here:

$$k = Ae^{-\frac{E_a}{RT}}$$

where A is a constant called the Arrhenius constant, E_a is the activation energy, R is the universal gas constant and T is the temperature in K.

Qualitative explanation of the Arrhenius equation

The value of A in the Arrhenius equation is almost constant over the ranges of temperatures for most reactions and since it appears in front of the exponential part of the equation it is often called the pre-exponential factor. It is also called the frequency factor for the reaction, as it is the total number of collisions leading to a reaction or not, per second.

The exponential part of the equation (i.e. $e^{-\frac{E_a}{RT}}$) is the probability that a given collision will result in a reaction. Either increasing the temperature or decreasing the activation energy (for example, by the use of catalysts) results in an increase in the value of k and hence in the rate of reaction.

Example

A reaction has the following rate equation:

$$rate = k\,[\mathrm{B}]$$

Explain qualitatively why increasing the temperature by 10 °C has a much greater effect on the rate of reaction for this reaction than doubling the concentration of B.

Answer

Reaction occurs when molecules have energy greater than or equal to the activation energy, E_a.

An increase in temperature by 10 °C will greatly increase the number of molecules with energies equal to or above the activation energy, E_a.

Doubling the concentration of B will only double the number of molecules with energies equal to or above the activation energy. Hence it only doubles the rate of reaction.

Rearranging the Arrhenius equation

The equation $k = Ae^{-\frac{E_a}{RT}}$ can be rearranged and manipulated into the form for the equation of a straight line (i.e. $y = mx + c$) in the following way:

Taking natural logarithms of both sides we obtain:

$$\ln k = \ln\left(Ae^{-\frac{E_a}{RT}}\right)$$

This can be separated to obtain.

$$\ln k = \ln A + \ln\left(e^{-\frac{E_a}{RT}}\right)$$

The natural logarithm of an exponential just leaves the power to which the exponential is raised. Hence we can write:

$$\ln k = \ln A - \frac{E_a}{RT}$$

Rearranging this equation so that we have it in the form $y = mx + c$, we obtain:

$$\ln k = -\frac{E_a}{RT} + \ln A$$

(Note that k and T are the two variables in this equation. All the other quantities (i.e. E_a, R, T and A) are constants.)

Comparing this equation to the equation for a straight line $y = mx + c$, if a graph of $\ln k$ is plotted on the y-axis with $\frac{1}{T}$ plotted on the x-axis, then the resulting graph will be a straight line with gradient of $-\frac{E_a}{R}$ and intercept on the y-axis of $\ln A$.

Examples

1 A reaction has a rate constant $k = 2.45 \times 10^{-2}$ mol^{-1} dm^{3} at 298 K.

Use the Arrhenius equation, $k = Ae^{-\frac{E_a}{RT}}$ to calculate a value for the activation energy for this reaction in kJ mol^{-1}.

The gas constant, $R = 8.31$ J K^{-1} mol^{-1}

The pre-exponential factor, $A = 2.51 \times 10^{10}$ mol^{-1} dm^{3}

Answer

1 (Note that the activation energy is in the power of the exponential, so we must take natural logarithms of both sides first.)

Taking natural logarithms of both sides we obtain:

$$\ln k = \ln\left(Ae^{-\frac{E_a}{RT}}\right)$$

This can be separated to obtain.

$$\ln k = \ln A + \ln\left(e^{-\frac{E_a}{RT}}\right)$$

The natural logarithm of an exponential just leaves the power to which the exponential is raised.
Hence we can write:

$$\ln k = \ln A - \frac{E_a}{RT}$$

Substituting in the numbers we have

$$\ln(2.45 \times 10^{-2}) = \ln(2.51 \times 10^{10}) - \frac{E_a}{8.31 \times 298}$$

$$-3.7091 = 23.9461 - \frac{E_a}{2476.38}$$

$$\frac{E_a}{2476.38} = 23.9461 + 3.7091$$

$$\frac{E_a}{2476.38} = 27.6552$$

$$E_a = 68\,485 \text{ J mol}^{-1}$$

Hence, $E_a = 68.5 \text{ kJ mol}^{-1}$

2 The following table shows values for the rate constant, k at different temperatures T.

Temp T /K	**Rate constant k /s^{-1}**
293	1.76×10^{-5}
308	1.35×10^{-4}
318	4.98×10^{-4}
338	4.87×10^{-3}

a) Use the values in the above table to complete the following table.

ln k	$\mathbf{\frac{1}{T}}$

b) Using the set of axes shown here and the values in the table in part (a)(i), plot a graph of $\ln k$ on the y-axis against $\frac{1}{T}$ on the x-axis.

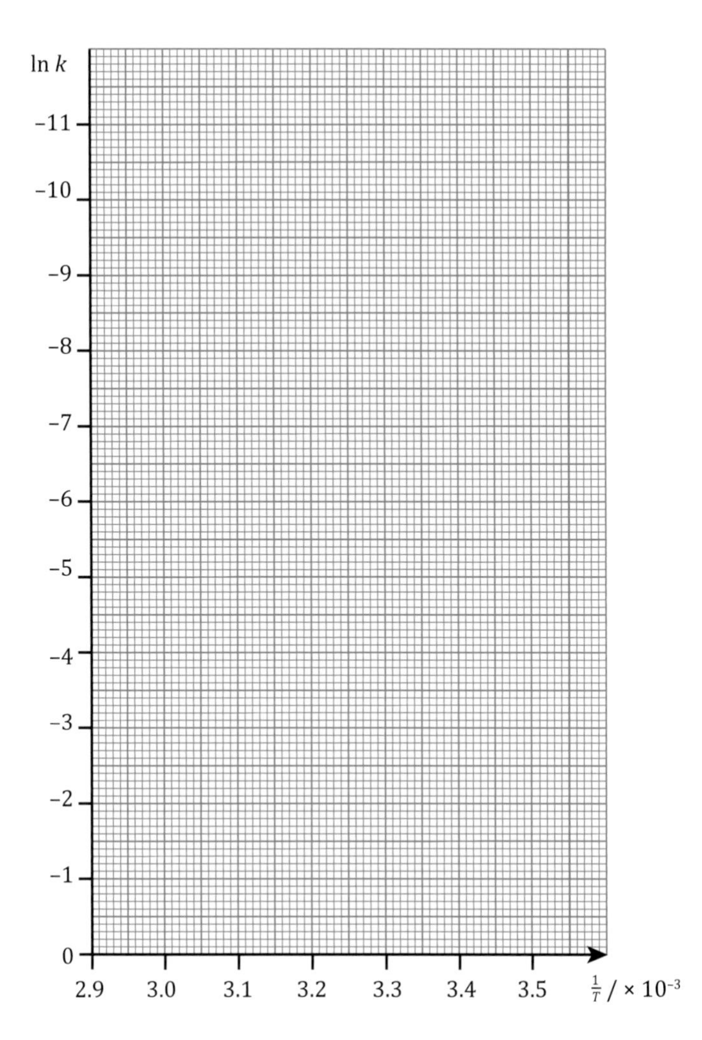

c) Find the gradient of the straight line and using the rearranged version of the Arrhenius equation in straight line form shown below to find a value for the activation energy, E_a.

$$\ln k = \ln A - \frac{E_a}{RT}$$

Answer

2 a)

ln *k*	$\frac{1}{T}$
−10.95	3.41×10^{-3}
−8.91	3.25×10^{-3}
−7.60	3.14×10^{-3}
−5.32	2.96×10^{-3}

b)

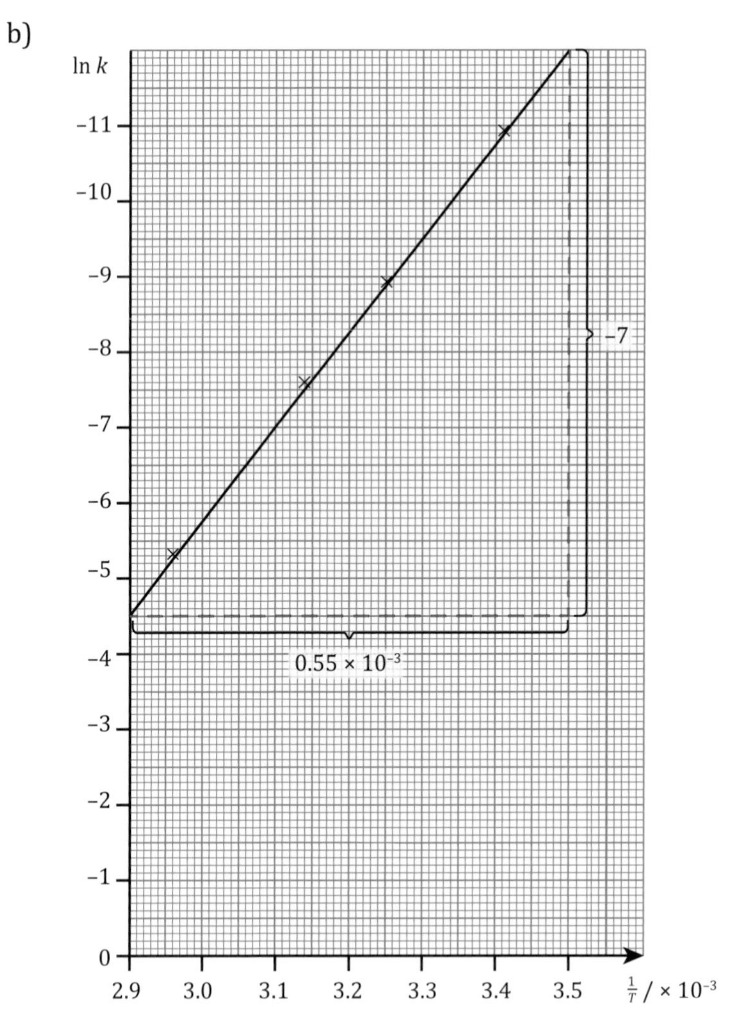

c) Gradient of line = $\dfrac{-7}{0.55 \times 10^{-3}}$ = −12 727.3

Now, from the equation for the straight line we have:

$$\text{Gradient} = -\frac{E_a}{R}$$

Hence $$-\frac{E_a}{R} = -12\,727.3$$

$$E_a = 12\,727.3 \times 8.31$$

$$= 105\,764\ \text{J mol}^{-1}$$

$$= 106\ \text{kJ mol}^{-1}$$

Test yourself 18

1. The rate equation for a certain reaction is:

   rate = $k[A][B]^3$

   (a) State the order of the reaction with respect to A.

   (b) State the order of the reaction with respect to B.

   (c) State the overall order for the reaction.

   (d) Explain what would happen to the initial rate if you doubled the concentration of:

   (i) A (ii) B

2. The equation for the reaction between hydrogen gas and iodine gas to form hydrogen iodide gas is shown here:

   $H_2(g) + I_2(g) \longrightarrow 2HI(g)$

   Experiments were conducted and the following information was found:

   The order of the reaction with respect to $[I_2] = 1$

   The order of the reaction with respect to $[H_2] = 1$

   The overall order for the reaction with respect to both reactants = 2

   From the information given, determine the rate equation.

3. Two gases A and B react together according to the following equation:

   $A(g) + 2B(g) \longrightarrow C(g) + D(g)$

   A series of experiments was conducted to find the initial rate of reaction at a constant temperature and the following rate equation was determined:

   rate = $k[A]^2[B]$

   (a) Complete the following table of data for the reaction between A and B:

Experiment	Initial concentration of A/mol dm^{-3}	Initial concentration of B/mol dm^{-3}	Initial rate/ mol dm^{-3}s^{-1}
1	1.5×10^{-2}	2.1×10^{-2}	4.0×10^{-5}
2	3.0×10^{-2}	2.1×10^{-2}	
3		4.2×10^{-2}	4.0×10^{-5}
4		6.3×10^{-2}	3.6×10^{-4}

   (b) Using the data from Experiment 1, calculate a value for the rate constant *k*, at this temperature and deduce the units for *k*.

4. Nitrogen monoxide and hydrogen react in the presence of a catalyst to give nitrogen and steam according to the following equation.

   $2NO(g) + 2H_2(g) \longrightarrow N_2(g) + 2H_2O(g)$

   Experiments were conducted and the rate equation was found to be:

   rate = $k[NO]^2[H_2]$

   If the initial concentration of NO was 1.5×10^{-2} mol dm^{-3} and the initial concentration of H_2 was 1.1×10^{-2} mol dm^{-3}, this resulted in a rate of reaction of 7.92×10^{-7} mol dm^{-3}s^{-1}.

   (a) Calculate the value of the rate constant, *k*, under these conditions and find its units.

   (b) If the initial concentration of H_2 was halved and the initial concentration of NO was doubled from their original values, calculate the initial rate of reaction.

5. The equation below shows the reaction between persulfate and iodide ions in aqueous solution:

   $S_2O_8^{2-} + 2I^- \longrightarrow 2SO_4^{2-} + I_2$

   To investigate the rate of this reaction, two experiments were carried out and the results are summarised in the following table:

Experiment	Initial concentration of $S_2O_8^{2-}$/mol dm^{-3}	Initial concentration of I^-/mol dm^{-3}	Initial rate/ mol dm^{-3}s^{-1}
1	0.0200	0.0100	4.00×10^{-6}
2	0.0400	0.0100	8.00×10^{-6}
3	0.0400	0.0200	1.60×10^{-5}

   (a) Determine the order of the reaction with respect to persulfate ions.

   (b) Determine the order of reaction with respect to iodide ions.

   (c) Write the rate equation for this reaction and use it to calculate the value of the rate constant, *k*, giving its units.

Chapter 19

The shapes of simple molecules

19.1 The shapes of simple molecules

There are a number of steps to take when working out what shape a molecule or ion will have and they are:

1. Decide which atom will be the central atom. The central atom appears once in the molecular formula. For example, in methane, CH_4, the central atom is carbon.
2. Determine the number of electrons in the outer level of the central atom. For example, in carbon, which is in Group 4, there will be 4 electrons in the outer level. Use the Group number, obtained from the periodic table, to guide you here. Be careful, though, as helium does not obey this rule.
3. Now add one electron for each of the surrounding atoms. For example, in methane there are 4 surrounding atoms (i.e. the hydrogen atoms) so we need to add 4 electrons to the 4 in the carbon outer shell. We now have 8 electrons to consider.
4. Divide by 2 to give the number of electron pairs. There are 8 electrons for methane which when divided by 2 gives 4. So 4 is the number of electron pairs for methane.
5. Decide how many bonding pairs and lone pairs there will be. As there are 4 surrounding atoms in methane and there are 4 pairs of electrons, these must all be bonding pairs. So, in methane there are no lone pairs (i.e. pairs of electrons which are not used for bonding).
6. Decide on the shape that will minimise the repulsion between the electron pairs (NB this is often a combination of bonding pairs and lone pairs of electrons). Electron pair repulsion theory is used for this and is covered in the next section.

19.2 Electron pair repulsion theory (also called valance shell electron pair repulsion (VSEPR) theory)

The shapes of simple molecules can be predicted using electron pair repulsion theory. Electron pairs repel each other, so they try to get as far from each other as possible to minimise the repulsion and this influences the shape of the molecule.

The strength of the repulsion depends on the type of pairs being considered according to the following diagram:

Smallest repulsion **Greatest repulsion**

Bonding pair — Bonding pair Lone pair — Bonding pair Lone pair — Lone pair

Repulsion increases

Pointer

You need to remember the order of repulsion and use it in questions where you are asked to explain shapes and bond angles.

Example

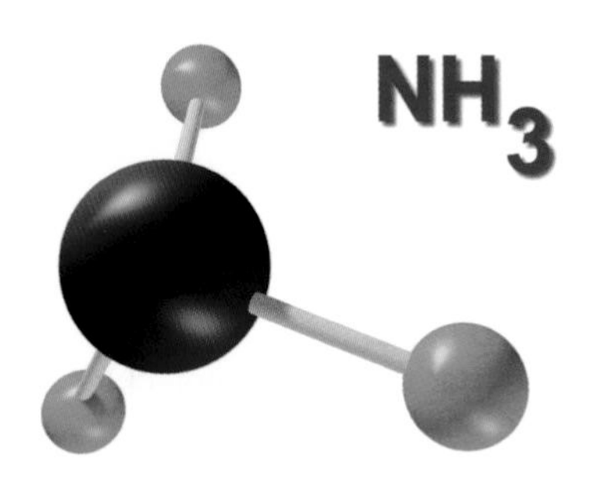

Draw the shape of an ammonia, NH_3, molecule.

Answer

The steps you need to take are outlined here but you do not need to write them down in questions like this. In many cases you will be able to do the working out in your head.

1. Nitrogen is the central atom.
2. Nitrogen is in Group 5, so there are 5 outer electrons.
3. Adding one electron each from the surrounding hydrogen atoms, gives 8 electrons.
4. Dividing by two gives 4 electron pairs
5. The three bonds to the hydrogen will take care of 3 electron pairs. There is one electron pair left which will be a lone pair. Hence there are 3 bonding pairs and one lone pair.

Electron pair repulsion theory is now used to predict the shape of NH_3.

There are two different types of repulsion to consider:

- Repulsion between the bonding pairs of electrons.
- Repulsion between the bonding pairs of electrons and the lone pair of electrons.

Note that according to the order of repulsions, the repulsion in the second bullet point is greater than that in the first.

There are four orbitals to consider with each containing an electron pair. The orbitals would adopt a tetrahedral arrangement with the N atom at the centre and the orbitals pointing towards the vertices of a tetrahedron.

The bond angle for a normal tetrahedral arrangement is 109.5°, but as there is greater repulsion between the lone pair and the bonding pairs, the bond angle between the N and H atoms is reduced to 107°.

» Pointer

Each orbital is the space occupied by an electron pair.

Now this is the difficult part to understand. We have just considered the arrangement of the orbitals and this is not necessarily the shape of the molecule. The lone pairs are not part of the shape for the actual molecule as they are not connected to another atom. Hence they are ignored when drawing the shape for the molecule.

Hence, the shape for NH_3 is pyramidal with the nitrogen atom at the top of the pyramid and the hydrogen atoms positioned at each of the remaining corners (called vertices). In most examination questions, you will be asked to mark on the diagram any lone pairs present.

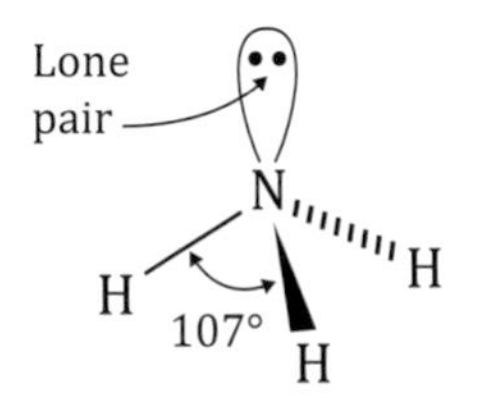

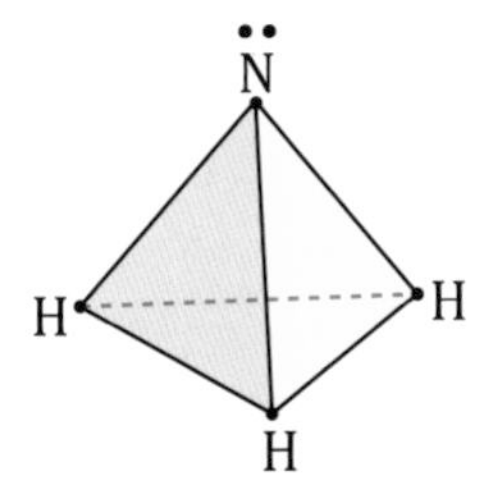

Pointer

You can see the pyramidal shape by joining up the atoms with lines like this. The lone pair can be marked simply as two dots.

The shapes of simple molecules can be determined once the numbers of bonding and lone pairs have been determined. You can see how this can be worked out for the following combinations.

Two bonding pairs of electrons

Repulsion from the two bonding pairs means the central atom is bonded to the two other atoms with a bond angle of 180°. This means that the molecule is linear (i.e. in a straight line).

Pointer
Note that this linear molecule has a bond angle of 180°.

An example of a molecule with this shape is $BeCl_2$:

Cl — Be — Cl

180°

Three bonding pairs of electrons

Repulsion between bonding pairs is equal so the molecule is a trigonal-planar shape with a bond angle of 120°.

Pointer
Remember that angles at a point add up to 360°.

Examples of molecules with three bonding pairs include BCl_3:

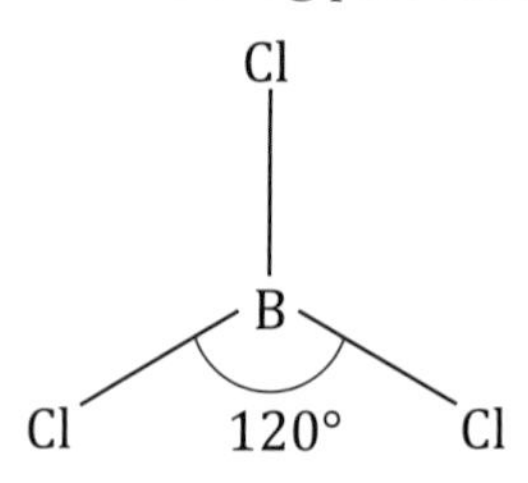

Four bonding pairs of electrons

The central atom has four bonding pairs with each bond pointing towards the vertices of a tetrahedron where the other atoms in the molecule are situated. The bond angles are 109.5°.

Examples include methane CH_4:

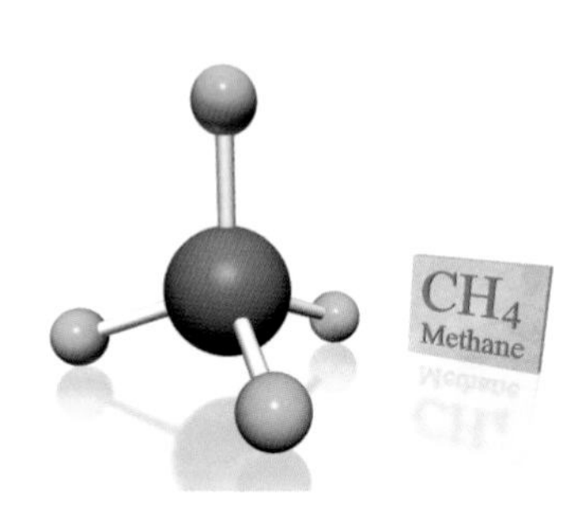

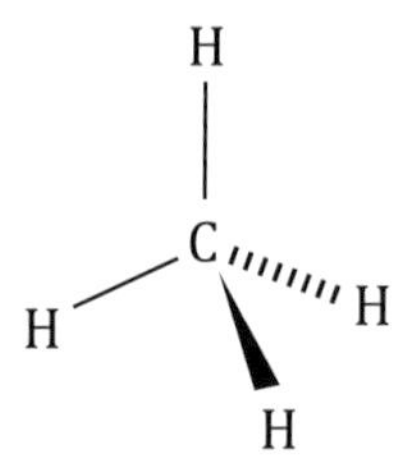

Five bonding pairs of electrons

The central atom has 3 bonding pairs in the same plane with bond angles of 120° and 2 bonding pairs at 90° to the plane with one above the plane and the other below the plane.

Pointer
Three of the chlorine atoms are in the same plane and their bond angle is 120°. Two more chlorine atoms lie above and below the central P atom at right-angles to the plane.

Examples include PCl_5:

Cl

Cl — P — Cl 120°

Cl

Cl

Six bonding pairs of electrons

The central atom has 4 bonding pairs in the same plane with bond angles of 90° and 2 bonding pairs at 90° to the plane with one above the plane and the other below the plane.

Examples include SF_6:

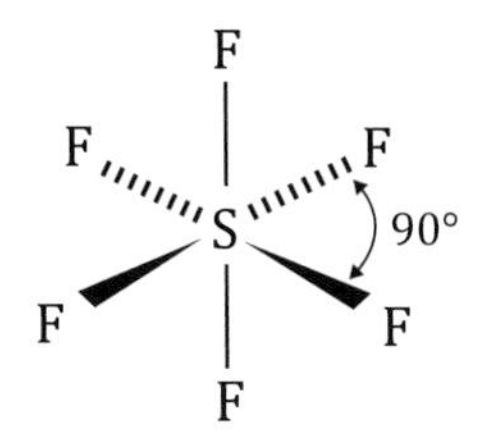

Note that the two F atoms above and below the S atom are at 90° above and below the plane containing the other F atoms

19.3 Simple molecules where there are both bonding pairs and lone pairs

Three bonding pairs and one lone pair

The 4 orbitals adopt a tetrahedral shape but as there are only 3 bonding pairs the shape of the molecule is pyramidal. Remember that the lone pair influences the shape but is not part of the shape. Because the lone pair pushes the bonding pairs down because the repulsion is greater, the bond angle will decrease to 107° (note for a tetrahedral arrangement the bond angle would be 109.5°).

Examples include NH_3.

Two bonding pairs and two lone pairs

The 4 orbitals adopt a tetrahedral shape but as there are only 2 bonding pairs the shape of the molecule is V-shaped or bent. The two lone pairs repel the bonding pairs more so this reduces the bond angle to 104.5°.

Examples include H_2O:

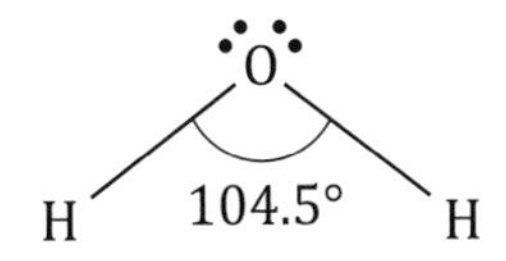

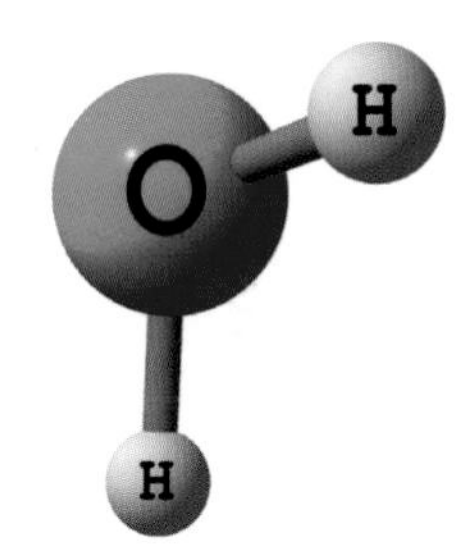

» *Pointer*
Bent molecules such as H_2O are sometimes called V-shaped molecules.

Example

Phosphorus is in Group 5 of the periodic table. Phosphorus reacts with hydrogen to form phosphine, PH_3.

a) Draw the shape of a phosphine molecule and include on your diagram any lone pairs of electrons.

b) Predict the bond angle in the phosphine molecule.

Answer

a) Phosphorus is in Group 5, so there are 5 electrons in the outer shell. The 3 hydrogen atoms will contribute 3 more electrons giving a total of 8 electrons or 4 electron pairs. As there are 3 bonds, 3 will be bonding pairs and one will be a lone pair. The orbitals will form a tetrahedral shape but the shape of the molecule will be trigonal-based pyramidal.

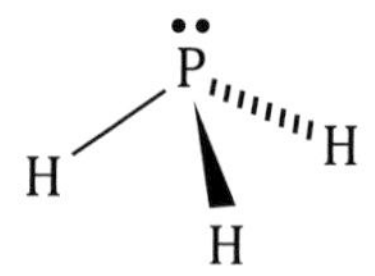

b) The bond angle will be 107°

Pointer
The lone pairs influence the shape but they are not part of the shape. You only consider the way the atoms are arranged.

Pointer
The repulsion between the lone-pair and bonding pair is greater than between the bonding pair and bonding pair. This reduces the angle to 107°.

19.4 The shapes of ions

An ion has a charge which can be positive or negative. An ion with a −1 charge has an extra electron added whereas an ion with a charge of +1 has lost an electron. We must take this into account when working out the total number of electrons.

This is best seen by looking at the following example.

Example

Ammonia reacts with a hydrogen ion to form the ammonium ion according to the following equation:

$$NH_3 + H^+ \longrightarrow NH_4^+$$

Predict the shape and the bond angle in a NH_4^+ ion.

Answer

Nitrogen is in Group 5 and has 5 electrons in the outer shell. The four hydrogen atoms will give another 4 electrons, making 9. However, the ion has a positive charge meaning that one electron has been lost. Hence there are a total of 8 electrons and this means there will be 4 electron pairs. As there are 4 bonds, the 4 pairs will all be bonding pairs.

There will be equal repulsion between the 4 bonding pairs of electrons so the ion will adopt a tetrahedral structure and the bond angle will be 109.5°.

Test yourself 19

1. Xenon, one of the noble gases, reacts with fluorine to give xenon tetrafluoride (XeF_4).
   (a) State the number of bonding pairs of electrons in an XeF_4 molecule.
   (b) State the number of lone pairs of electrons in an XeF_4 molecule.
   (c) Draw a diagram showing the shape of the XeF_4 molecule. Mark on your diagram the position of any lone pairs.
   (d) Give the name of the shape of an XeF_4 molecule.

2. A phosphine molecule (PH_3) reacts with a hydrogen ion (H^+) to form a PH_4^+ ion.
   Predict the shape and the bond angle in a PH_4^+ ion.
3. Fluorine reacts with many elements and will even react with other halogens.
   When molecules of bromine trifluoride (BrF_3) react together, the following two ions are produced

   BrF_4^- and BrF_2^+

   Draw the shapes of each of these ions and for each one, predict the bond angle.

Chapter 20 The solubility of compounds in water

20.1 The solubility of compounds in water

When a compound is dissolved in water, the compound being dissolved is called the solute and the water is called the solvent.

The solubility of a substance at a given temperature is the mass of the substance that will dissolve in a given mass of solvent to form a saturated solution at that temperature.

The units of solubility are grams of solute per given mass of solvent, e.g. g per 100 g of solvent. Solubility may also be expressed as moles of solute per given mass of solvent, e.g. mol kg^{-1}.

Dilute solutions, saturated solutions and supersaturated solutions

The solubility of a substance is the upper limit of the concentration of the solute. It is the maximum amount of solute that can be dissolved in a certain mass of solvent.

When a solid is added to water there are a number of things that can happen:

- If the solution contains less solute than the maximum amount that the mass of water could contain, then the solute will dissolve completely to produce a dilute solution.
- If the amount of solute being dissolved matches exactly the maximum amount of solute the water is capable of dissolving, then a saturated solution is produced.
- If the amount of solute being dissolved exceeds the maximum amount the water is capable of dissolving, then the excess solute will be left undissolved.
- If the saturated solution and the undissolved solute were heated, the solute would dissolve. Now if you then cooled the solution down, it is possible for the solute to stay in the solution. When this happens, the solution is said to be supersaturated and such a solution is unstable and if you add a seed (such as a very small crystal) the extra solute will crystallise out and the solution will go back to being saturated.

The solubility of a solute usually decreases with decreasing temperature, so as the temperature decreases some of the solute will appear (sometimes as crystals) and the mass of the solute which is undissolved can be determined by subtracting the solubility at the lower temperature from the solubility at the higher temperature. The following example shows this.

Examples

1 Barium chloride is soluble in water. The solubility of barium chloride at two different temperatures is shown in the following table:

Temperature/°C	Solubility of $BaCl_2$/ g dm^{-3}
0	312
20	358

Calculate the mass of solid barium chloride that would be produced if 250 cm^3 of a saturated solution was cooled from 20 °C to 0 °C.

Answer

1 At 0 °C a 250 cm^3 saturated solution would contain

$\frac{312}{4} = 78$ g of $BaCl_2$

At 20 °C a 250 cm^3 saturated solution would contain

$\frac{358}{4} = 89.5$ g of $BaCl_2$

Hence, mass produced = 89.5 − 78 = 11.5 g

» Pointer

Note that the solubility in the table is expressed in g dm^{-3}. 1 dm^3 = 1000 cm^3. As there is only 250 cm^3 of solution, we divide the solubilities by 4.

2 The graph below shows the variation of the solubility of potassium chlorate ($KClO_3$) with temperature:

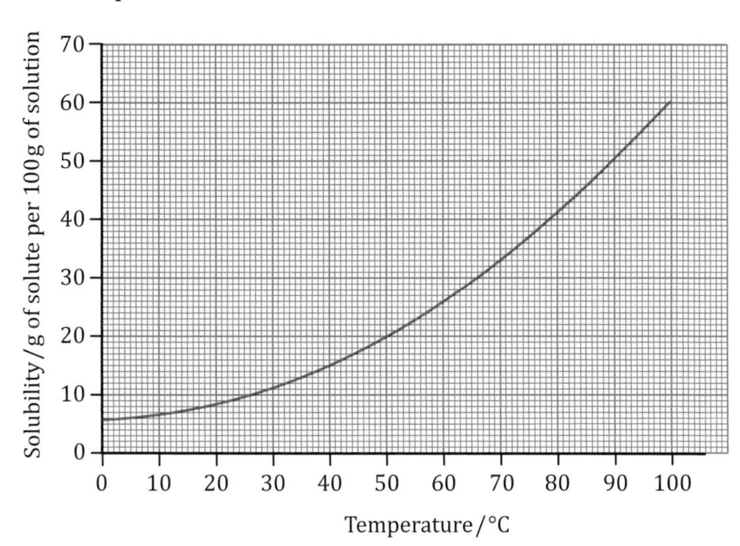

a) Define the term solubility.

b) Using the graph, say whether the following solutions are dilute (unsaturated), saturated or supersaturated:

i) 35g of $KClO_3$ in 100g of solution at 80 °C.

ii) 70g of $KClO_3$ in 100g of solution at 90 °C.

iii) 20g of $KClO_3$ in 100g of solution at 50 °C.

Answer

2 a) The solubility of a substance at a given temperature is the mass of the substance that will dissolve in a given mass of solvent to form a saturated solution at that temperature.

b) i) From the graph, at 80 °C, 40g of $KClO_3$ would produce a saturated solution.

35 g would therefore produce an unsaturated/dilute solution.

ii) From the graph, at 90 °C, 50g of $KClO_3$ would produce a saturated solution.

70 g would therefore produce a supersaturated solution.

iii) From the graph at 50 °C, 20g of $KClO_3$ would produce a saturated solution so as this is exactly the amount added, the solution produced will be saturated.

Test yourself 20

1. The graph below shows how the solubility of a solute in water varies with temperature:

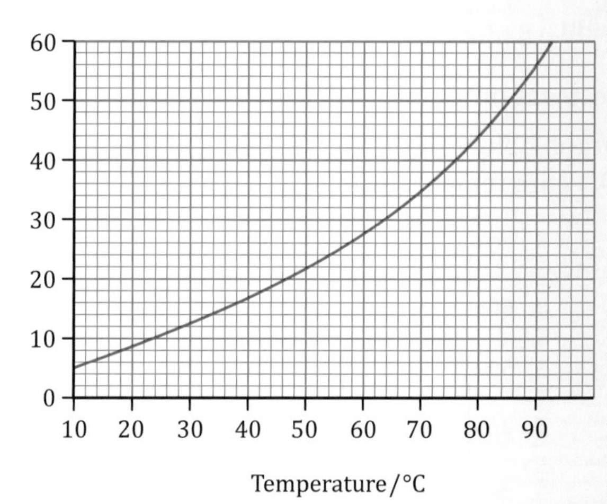

(a) Calculate the solubility of the solute in 100 g of solution at 20 °C.

(b) Use the solubility curve to determine the maximum mass of solid that would form from 100 g of solution cooled from 70 °C to 30 °C.

2. A compound X is soluble in water. The solubility of compound X at two different temperatures is shown in the following table:

Temperature/°C	Solubility of X/ g dm^{-3}
10	45.0
30	64.0

Calculate the mass of solid X that would be produced if 200 cm^3 of a saturated solution was cooled from 30 °C to 10 °C.

Chapter 21 Oxidation states (or numbers)

21.1 Oxidation and reduction

Oxidation is loss of electrons and reduction is gain of electrons. We can remember this using the following:

O I L	**R I G**
Oxidation is loss	Reduction is gain

A reaction in which oxidation and reduction both take place is called a redox reaction.

Take the following reaction, for example:

$$2Ca + O_2 \longrightarrow 2CaO$$

To understand what is happening we can split the equation into two, called half-equations. Using the half-equations we can see which of the reactants gains electrons and which loses electrons.

The half-equation for the calcium is as follows:

$$2Ca \longrightarrow 2Ca^{2+} + 4e^-$$

The half-equation for the oxygen is as follows:

$$O_2 + 4e^- \longrightarrow 2O^{2-}$$

If you now look at the full reaction (i.e. $2Ca + O_2 \longrightarrow 2CaO$) you can see that it represents a redox reaction as both oxidation and reduction take place. The calcium is being oxidised and the oxygen is being reduced. The species bringing about the oxidation (i.e., oxygen) is called the oxidising agent and the species bringing about the reduction (i.e. the calcium) is called the reducing agent.

» Pointer

The calcium has lost electrons to form calcium ions. Electron loss is oxidation (remember OIL). Use the full equation to determine the number of calcium atoms that react (i.e. 2 in this case).

» Pointer

The oxygen has gained electrons to form oxide ions. Electron gain is reduction (remember RIG).

21.2 Oxidation states (also called oxidation numbers)

The oxidation state or number of an atom is the number of electrons used by the atom when it bonds with other atoms of other elements. We use oxidation states or numbers to determine whether a particular atom has been oxidised or reduced during a reaction.

There are a number of rules to follow:

- If the atom is uncombined (i.e. not bonded to atoms of other elements), its oxidation state is 0. It is important to note that if it is bonded to atoms of the same element, e.g. O_2, then the oxidation state is still 0.
- The sum of all the oxidation states of all the atoms or ions in a neutral compound is zero.
- In any ion, the sum of oxidation states of the atoms making up the ion, equals the charge on the ion.

Here is a table which shows the oxidation states (numbers) of common elements you may find in questions:

Element	Oxidation state	Notes about exceptions
Uncombined elements (e.g. Na, O_2, S)	0	
Combined Group 1 metals (e.g. Na, K, etc.)	+1	
Combined Group 2 metals (e.g. Ca, Mg)	+2	
Combined halogens (e.g. F, Cl, etc.)	−1	
Combined oxygen	−2	Does not apply if the oxygen is bonded to fluorine or the oxygen is in a peroxide.
Combined hydrogen	+1	For metal hydrides when hydrogen is bonded with a Group I metal (e.g. NaH, KH), the oxidation state is −1.

The following examples show how the oxidation states (numbers) in the above table can be used to determine the oxidation state (number) of an element in a compound.

Examples

1 Find the oxidation state of S in the compound sulfuric acid (H_2SO_4).

Answer

1 Each O has an oxidation state of −2, so as there are 4 of these, this will give a contribution of −8.

Each H has an oxidation state of +1, so two of these will give a contribution of +2.

Adding the two contributions together gives −8 + 2 = −6.

As the compound is neutral, the sulfur must have an oxidation state of +6 (because − 6 + 6 = 0)

2 Determine the oxidation state of Mn in Mn_2O_7.

Answer

2 O has an oxidation state of −2 and as there are 7 O atoms this gives −14.

The two Mn atoms have to have an oxidation state of +14 to give zero when added to the −14. Hence the oxidation state of one of the Mn atoms is +7.

Oxidation number of Mn is +7

3 Determine the oxidation number of iron in the ion $Fe(CN)_6^{4-}$, if the cyanide ion has an oxidation number of −1.

Answer

3 The cyanide ion has an oxidation number of −1. Six of these will give a total of −6. Now as the ion has a charge of −4 we need to add a positive number to −6 to give −4. That positive number is +2, which is the oxidation number of Fe.

21.3 Determining whether a reactant has been oxidised or reduced

Oxidation states can be used to determine whether a reactant has been oxidised or reduced during a redox reaction.

When an element is oxidised, its oxidation state increases and when it is reduced its oxidation state decreases.

This will be covered in the next chapter on redox reactions.

Test yourself 21

1. Determine the oxidation state of the element underlined in each of the following compounds:
   (a) $\underline{C}O_2$
   (b) $K_2\underline{S}O_4$
   (c) $\underline{Na}_2O$
   (d) $Mg\underline{C}O_3$
   (e) $\underline{Fe}_2O_3$
2. Determine the oxidation number of each of the named elements in the following compounds or ions:
   (a) Cr in CrO_3
   (b) Cu in CuO
   (c) C in CO_3^{2-}
   (d) Fe in $Fe(CN)_6^{3-}$
   (e) Cl in ClO^-
   (f) Br in BrO_3^-
   (g) S in SO_3^{2-}
   (h) N in N_2O_7
   (i) I in IO_4^-
   (j) Cr in $Cr_2O_7^{2-}$
3. Calcium manganate $Ca(MnO_4)_2$ is often used as a catalyst.

   Determine the oxidation state (number) of manganese in calcium manganate.
4. Sodium chlorate is often used as a weed killer to keep paths and drives free from weeds. The formula for sodium chlorate is $NaClO_3$.

   Determine the oxidation state of Cl in $NaClO_3$.
5. Sodium reacts with chlorine to form sodium chloride:

   $$2Na + Cl_2 \longrightarrow 2NaCl$$

   The above reaction is a redox reaction.

   Draw two half-equations for the above reaction, one to represent the reduction and the other to represent the oxidation.

Chapter 22 Redox reactions

22.1 Redox reactions

Redox stands for reduction–oxidation. Redox reactions are those reactions where atoms have their oxidation states/numbers changed. Electrons are transferred from one species to another with one species losing electrons and the other species gaining them. In the last chapter we used OIL RIG to help remember that oxidation is loss of electrons and reduction is gain of electrons.

Example

Magnesium reacts with chlorine to form magnesium chloride:

$$Mg + Cl_2 \longrightarrow MgCl_2$$

a) Write separate half-equations for the above reaction.

b) Using the half-equations, determine which species is oxidised and which is reduced.

c) Identify the oxidising agent and the reducing agent in the above reaction.

Answer

a) $Mg \longrightarrow Mg^{2+} + 2e^-$

$Cl_2 + 2e^- \longrightarrow 2Cl^-$

b) $Mg \longrightarrow Mg^{2+} + 2e^-$ The Mg has lost electrons so it has been oxidised.

$Cl_2 + 2e^- \longrightarrow 2Cl^-$ The Cl_2 has gained electrons so it has been reduced.

c) The oxidising agent is the species that is itself reduced. Hence, Cl_2 is the oxidising agent.

The reducing agent is the species that is itself oxidised. Hence, Mg is the reducing agent.

» Pointer
You could use oxidation states to work this out or use your knowledge about the charges on the ions.

» Pointer
Use OIL RIG to help remember that oxidation is loss of electrons whilst gain of electrons is reduction.

Using oxidation states in redox reactions

Oxidation states/numbers can be used to determine whether a particular atom has been oxidised or reduced in a reaction. If there has been an increase in oxidation number, the atom has been oxidised and if there is a decrease then it will have been reduced.

Take the following reaction:

$$3Cu + 8H^+ + 2NO_3^- \longrightarrow 3Cu^{2+} + 2NO + 4H_2O$$

By obtaining the oxidation states of the nitrogen in NO_3^- and NO we can decide whether it has been oxidised or reduced.

The oxidation state of N in NO_3^- is obtained in the following way. The oxidation of combined oxygen is −2 and since there are three atoms this will give a total of −6. Now since NO_3^- is an ion with a −1 charge, the oxidation state of N added to

−6 must give the overall charge on the ion (i.e. −1). Now −6 + 5 = −1, so N has an oxidation state of +5.

For NO the oxidation state of the N is +2 (i.e. to balance the −2 of the oxygen).

In the reaction shown, the oxidation state of N has changed from +5 to +2. This is a decrease in oxidation state which indicates that the N has been reduced.

If we consider the Cu which goes from oxidation state 0 (because it is an uncombined element) to oxidation state +2, this is an increase in oxidation state so the copper has been oxidised.

22.2 Proving that a reaction is a redox reaction

To prove that a reaction is a redox reaction you need to show that reduction and oxidation occur in the same reaction. This can be done by considering the change in the oxidation states of some of the species in the reaction as the following example shows.

Example

Hydrazine can be used as rocket fuel. Hydrazine can be produced from ammonia and sodium hypochlorite (NaClO).

$$2NH_3 + NaClO \longrightarrow N_2H_4 + NaCl + H_2O$$

Using oxidation numbers, explain why this reaction is a redox reaction.

Answer

In NH_3 the oxidation state of hydrogen is +1 so as there are three atoms, this gives +3. Hence the oxidation state of N in NH_3 is −3.

The oxidation states in a molecule must add up to zero.

In N_2H_4 the oxidation state of the 4 hydrogens is +4. Hence the oxidation state of the two nitrogens is −4. Hence the oxidation state of N in N_2H_4 is −2.

The oxidation state of N has increased from −3 to −2. This represents oxidation.

In NaClO, the oxidation states of Na and O are +1 and −2 respectively. This gives −1 so the Cl must have an oxidation state of +1.

In NaCl the oxidation state of Cl is −1.

The oxidation state of Cl has decreased from +1 to −1 so this represents reduction.

As reduction and oxidation occurs in the same reaction, it is a redox reaction.

22.3 Proving that a reaction is not a redox reaction

To prove that a reaction is not a redox reaction you simply show that the oxidation state of the element or species being considered has not changed during the reaction. Hence the oxidation state of the atom in the reactants will be the same as that in the products.

Example

Ozone is formed in the upper atmosphere by the action of ultra violet radiation on oxygen molecules according to the following equation:

$$3O_2 \longrightarrow 2O_3$$

Show that the reaction shown by the above equation is not a redox reaction.

Answer

The oxygen in the oxygen molecules in the reactants is uncombined with other elements so the oxidation state is zero. The oxygen in the products (i.e. ozone) is also uncombined and therefore has an oxidation state of zero.

Since there is no change in the oxidation state of oxygen, this is not a redox reaction.

Test yourself 22

1. Here is an equation for a redox reaction:

   $$2NO + 12H^+ + 10I^- \longrightarrow 2NH_4^+ + 5I_2 + 2H_2O$$

   (a) Explain what is meant by a redox reaction.

   (b) Find the oxidation states of nitrogen in NO and NH_4^+ and hence state whether the nitrogen in this reaction has been oxidised or reduced.

2. (a) Explain in terms of the electrons, what happens to an oxidising agent during a redox reaction.

   (b) Zinc reacts with hydrochloric acid to give zinc chloride and hydrogen gas according to the following equation:

   $$Zn(s) + 2HCl(aq) \longrightarrow ZnCl_2(aq) + H_2(g)$$

   (i) State the oxidising agent in the above reaction.

   (ii) State the oxidation number of zinc in Zn(s) and $ZnCl_2(aq)$.

   (iii) Hence determine whether the zinc is oxidised or reduced in the reaction.

3. Ammonia can be used to produce nitric acid. In the first part of the process, ammonia and air are passed over a catalyst at 850 °C.

   The equation for this reaction is:

   $$4NH_3(g) + 5O_2(g) \longrightarrow 4NO(g) + 6H_2O(g)$$

   (a) Complete the table below, giving the oxidation states/numbers of each element present.

Element	**Initial oxidation state**	**Final oxidation state**
nitrogen		
hydrogen		
oxygen		

   (b) Use the completed table to explain which species have been oxidised in this reaction.

Chapter 23 Electrochemical cells

Electrochemical cells produce electricity from chemical reactions and they form the basis of batteries. If the cell reaction can be reversed then the battery can be made re-chargeable.

23.1 Electrodes and half-cells

When a metal rod is placed in a solution of its own ions, an equilibrium is set up. For example, the equilibrium for an Mg rod in a solution of magnesium sulfate (i.e. a source of Mg^{2+} ions) is:

$$Mg(s) \rightleftharpoons Mg^{2+}(aq) + 2e^-$$

This equilibrium is far over to the right so magnesium atoms in the rod will lose electrons and the resulting ions will go into solution as Mg^{2+} ions. The electrons will remain in the rod which will leave it with a negative charge.

Connecting two electrodes together

A metal in a solution of its own ions is called an electrode. When two different electrodes are connected by wires and a salt bridge to a voltmeter, then a potential difference (i.e. voltage) is produced between the two wires which can be measured by the voltmeter. The purpose of the salt bridge is to provide a path for the electrons from one solution to the other. Wires cannot be used for this because they could set up their own potential difference when placed in the solutions of ions.

Suppose a zinc electrode and a copper electrode are connected together as shown in the diagram below:

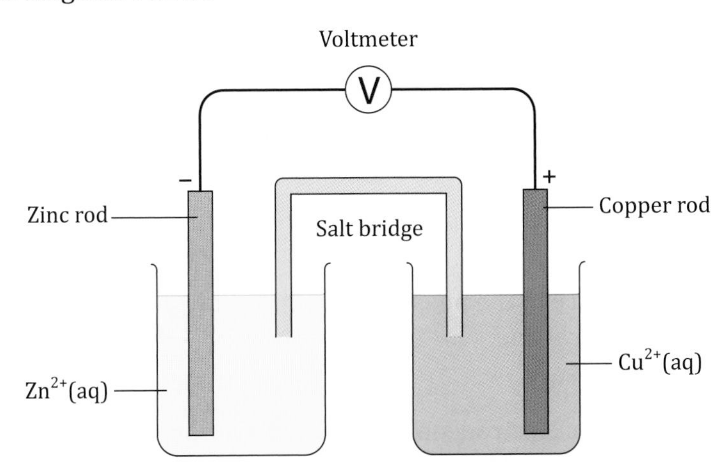

The following reactions take place at each electrode:

$$Zn(s) \longrightarrow Zn^{2+}(aq) + 2e^-$$

Pointer
This is the half-equation showing that at the Zn electrode, electrons are lost. Hence the zinc is oxidised. Remember that OIL in OIL RIG means oxidation is loss of electrons.

$$Cu^{2+}(aq) + 2e^- \longrightarrow Cu(s)$$

This is the half-equation showing that at the Cu electrode, electrons are gained. Hence the Cu^{2+} is reduced. Remember that the RIG in OIL RIG means that reduction is gain of electrons.

Adding the two half-equations together we obtain:

$$Zn(s) + Cu^{2+}(aq) \longrightarrow Zn^{2+}(aq) + Cu(s)$$

Notice the electrons do not appear in the full equation. They can be cancelled as the same numbers of electrons appear on both sides of the equation.

As each electrode is at a different potential, a potential difference called the cell emf is produced which can be measured by a voltmeter. The cell emf can be calculated if the standard electrode potential of each electrode is known. The following section explains how this is done.

Using standard electrode potential data

Each half-equation has potential associated with it and this can be used to work out the potential difference when two different electrodes are connected together.

Here is a table of electrode half-equations and their corresponding standard electrode potentials:

Electrode half-equation	**Standard electrode potential $E^\ominus$/V**
$H_2O_2(aq) + 2H^+(aq) + 2e^- \longrightarrow 2H_2O(l)$	+1.77
$Cl_2(g) + 2e^- \longrightarrow 2Cl^-(aq)$	+1.36
$O_2(g) + 4H^+(aq) + 4e^- \longrightarrow 2H_2O(l)$	+1.23
$Ag^+(aq) + e^- \longrightarrow Ag(s)$	+0.80
$Fe^{3+}(aq) + e^- \longrightarrow Fe^{2+}(aq)$	+0.77
$Cu^{2+}(aq) + 2e^- \longrightarrow Cu(s)$	+0.34
$Fe^{2+}(aq) + 2e^- \longrightarrow Fe(s)$	−0.44
$Zn^{2+}(aq) + 2e^- \longrightarrow Zn(s)$	−0.76
$Al^{3+}(aq) + 3e^- \longrightarrow Al(s)$	−1.66

Reducing agents bring about reduction and are themselves oxidised as a result.

Oxidising agents bring about oxidation and are themselves reduced as a result.

Pointer

The reaction $Cu^{2+}(aq) + 2e- \longrightarrow Cu(s)$ has $E^\ominus = +0.34V$. The reaction $2Cu^{2+}(aq) + 4e- \longrightarrow 2Cu(s)$ would still have $E^\ominus = +0.34V$ as $E^\ominus$ does not depend on the number of moles.

There are a number of important things to note about this table:

- Good reducing agents (i.e. species that lose electrons easily) are at the bottom of the table (more negative values of $E^\ominus$).
- Good oxidising agents are at the top of the table (more positive values of $E^\ominus$)
- The electrons are on the left of the half-equation. If the reaction is reversed, then the sign of $E^\ominus$ will need to be reversed.

The number of electrons involved in the half-equation has no bearing on the value of $E^\ominus$.

Connecting two standard electrodes together

The following shorthand can be used for writing down the cell formed by two electrodes.

| represents a difference in the phase (i.e. solid, liquid, gas) between the species either side of the vertical line

|| represents a salt bridge

Suppose we have a zinc electrode in a solution of zinc sulfate (i.e. a source of Zn^{2+} ions) connected via a salt bridge to a copper electrode in a solution of copper sulfate (i.e. a solution of Cu^{2+} ions).

The standard electrode potentials for the following reactions are given or can be looked up:

$$Zn^{2+}(aq) + 2e^- \longrightarrow Zn(s) \qquad E^\ominus = -0.76\ V$$

$$Cu^{2+}(aq) + 2e^- \longrightarrow Cu(s) \qquad E^\ominus = +0.34\ V$$

The standard electrode potential of copper (i.e. +0.34 V) is higher than that of zinc (i.e. −0.76 V), so the electrons will flow from the zinc to the copper. Note that electrons always flow from the more negative to the more positive potential. When writing the shorthand representation for the cell we write the more positive electrode to the right.

Hence, the shorthand representation for the cell is

$$Zn(s)\,|\,Zn^{2+}(aq)\,||\,Cu^{2+}(aq)\,|\,Cu(s)$$

To work out $E^\ominus_{cell}$ we use the following formula:

$$E^\ominus_{cell} = E^\ominus_{\text{right-hand electrode}} - E^\ominus_{\text{left-hand electrode}}$$

For this electrochemical cell

$$E^\ominus_{cell} = E^\ominus_{\text{right-hand electrode}} - E^\ominus_{\text{left-hand electrode}} = +0.34 - (-0.76)$$
$$= +0.34 + 0.76$$
$$= +1.10\ V$$

Examples

1 The cell represented below was set up:

$$Ni(s)\,|\,Ni^{2+}(aq)\,||\,Ag^{+}(aq)\,|\,Ag(s)$$

The standard electrode potentials are:

$$Ni^{2+}(aq) + 2e^- \longrightarrow Ni(s) \qquad E^\ominus = -0.26\ V$$

$$Ag^{+}(aq) + e^- \longrightarrow Ag(s) \qquad E^\ominus = +0.80\ V$$

Deduce the emf of the cell.

Answer

1 $E^\ominus_{cell} = E^\ominus_{\text{right-hand electrode}} - E^\ominus_{\text{left-hand electrode}}$

$= +0.80 - (-0.26)$

$= +0.80 + 0.26$

$= +1.06\ V$

2 Here are two standard half-cells and their associated standard electrode potentials:

$$Pb^{2+}(aq) + 2e^- \longrightarrow Pb(s) \qquad E^\ominus = -0.13\ V$$

$$Ni^{2+}(aq) + 2e^- \longrightarrow Ni(s) \qquad E^\ominus = -0.23\ V$$

a) Give the cell reaction that occurs when the half-cells are connected to form a battery.

b) Calculate the emf of the battery.

Answer

2 a) The standard electrode potential of lead (i.e. −0.13 V) is higher than that of nickel (i.e. −0.23 V), so the electrons will flow from the nickel to the lead.

Hence electrons are produced at the nickel electrode. The half-reactions are:

» *Pointer*

Looking at this shorthand you can tell the following. The zinc is oxidised to form Zn^{2+} ions, whilst the Cu^{2+} ions are reduced to Cu. We can also tell that the Cu is the positive electrode and the Zn is the negative electrode as the positive electrode is on the right.

» *Pointer*

When two minuses are next to each other like this −(−0.76) the sign changes to a plus. If you are unsure about this you should review the material on directed numbers on page 8.

» *Pointer*

The right-hand cell electrode will be the one with the higher standard electrode potential. In this case it will be the Ag electrode.

$Ni(s) \longrightarrow Ni^{2+}(aq) + 2e^-$

$Pb^{2+}(aq) + 2e^- \longrightarrow Pb(s)$

Adding these two equations gives the overall reaction:

$Ni(s) + Pb^{2+}(aq) \longrightarrow Ni^{2+}(aq) + Pb(s)$

b) $E^\ominus_{cell} = E^\ominus_{\text{right-hand electrode}} - E^\ominus_{\text{left-hand electrode}}$

$= -0.13 - (-0.23)$

$= -0.13 + 0.23$

$= 0.10\ V$

Pointer

The right-hand electrode is always the electrode with the more positive (or less negative) standard electrode potential. As −0.13 is more positive than −0.23, the lead electrode will be the right-hand electrode.

23.2 Using standard electrode potentials to determine whether or not a redox reaction will take place

To determine whether a given redox reaction is feasible or not, the emf of the cell based upon the given redox reaction is worked out using the standard electrode potentials obtained from the earlier table. For a redox reaction to be spontaneous, the emf of the cell must be positive. If the emf is negative then the reaction will not be spontaneous.

Examples

1 Use the following half-equations and their standard electrode potentials to work out if magnesium powder would reduce copper(II) ions in solution to copper:

$Cu^{2+}(aq) + 2e^- \longrightarrow Cu(s)$ $\quad E^\ominus = +0.34\ V$

$Mg^{2+}(aq) + 2e^- \longrightarrow Mg(s)$ $\quad E^\ominus = -2.37\ V$

Answer

1 The standard electrode potential of copper (i.e. +0.34 V) is higher than that of magnesium (i.e. −2.37 V), so the electrons will flow from the magnesium to the copper.

Hence electrons are produced at the magnesium electrode so the half-reaction here is:

$Mg(s) \longrightarrow Mg^{2+}(aq) + 2e^-$

The half-reaction at the copper electrode will be:

$Cu^{2+}(aq) + 2e^- \longrightarrow Cu(s)$

The magnesium has lost electrons and so has been oxidised to $Mg^{2+}(aq)$ whilst the $Cu^{2+}(aq)$ has gained electrons so it has been reduced to Cu(s).

Hence magnesium powder would reduce copper(II) ions in solution to copper.

Alternative answer

1 An alternative answer would be to assume that magnesium powder would reduce copper(II) ions in solution to copper and then determine the emf for the cell. If the emf is positive, then the reaction is feasible.

The standard electrode potential of copper (i.e. +0.34 V) is higher than that of magnesium (i.e. −2.37 V), so the electrons will flow from the magnesium to the copper. The more positive electrode (i.e. the copper) will be the right-hand electrode. The shorthand for this cell is:

$Mg(s) \mid Mg^{2+}(aq) \parallel Cu^{2+}(aq) \mid Cu(s)$

$$E^{\ominus}_{cell} = E^{\ominus}_{\text{right-hand electrode}} - E^{\ominus}_{\text{left-hand electrode}}$$
$$= +0.34 - (-2.37)$$
$$= +0.34 + 2.37$$
$$= +2.71\ V$$

As $E^{\ominus}_{cell}$ has a positive value, the reaction is feasible.

2 When powdered copper is added to an acidified solution of potassium manganate(VII), a redox reaction takes place.

The half-equations for this and their standard electrode potentials are as follows:

$Cu^{2+}(aq) + 2e^- \longrightarrow Cu(s)$ $\quad E^{\ominus} = +0.34\ V$

$MnO_4^-(aq) + 8H^+ + 5e^- \longrightarrow Mn^{2+}(aq) + 4H_2O(l)$ $\quad E^{\ominus} = +1.52\ V$

Will the copper reduce the manganate(VII) ions ($MnO_4^-(aq)$) to manganese(II) ions?

Answer

2 From the values of the standard electrode potentials $E^{\ominus}$, electrons will flow from the more negative electrode to the more positive electrode. As +0.34 V has a lower value than +1.52 V, the electrons will flow from the Cu(s) to the potassium manganate(VII). Hence the copper will be oxidised to $Cu^{2+}(aq)$. The MnO_4^- (i.e. a source of Mn(VII)) will gain electrons and will therefore be reduced to $Mn^{2+}(aq)$ (i.e. a source of Mn(II)).

Hence the copper will reduce the manganate(VII) ions ($MnO_4^-(aq)$) to manganese (II) ions.

3 Using the following half-equations and standard electrode potentials, determine whether it is possible for copper to reduce nickel(II) ions in solution to nickel:

$Ni^{2+}(aq) + 2e^- \longrightarrow Ni(s)$ $\quad E^{\ominus} = -0.25\ V$

$Cu^{2+}(aq) + 2e^- \longrightarrow Cu(s)$ $\quad E^{\ominus} = +0.34\ V$

Answer

3 The standard electrode potential of copper (i.e. +0.34 V) is higher than that of nickel (i.e. −0.25 V), so electrons will flow from the nickel to the copper. This means that the nickel is oxidised to Ni^{2+} and the copper is reduced. Hence it is not possible to reduce nickel(II) ions in solution to nickel.

Another way to answer this is to find the sign of $E^{\ominus}$ for the cell.

If copper reduced nickel(II) ions in solution to nickel the copper would change to Cu^{2+} releasing electrons which would then reduce the Ni^{2+} to Ni. The shorthand for this cell would be as follows:

$Cu(s) \mid Cu^{2+}(aq) \parallel Ni^{2+}(aq) \mid Ni(s)$

$$E^{\ominus}_{cell} = E^{\ominus}_{\text{right-hand electrode}} - E^{\ominus}_{\text{left-hand electrode}}$$
$$= -0.25 - (+0.34)$$

» Pointer

The right-hand electrode is always the electrode with the more positive (or less negative) standard electrode potential. As +0.34V is more positive than −2.37V, the copper electrode will be the right-hand electrode.

$= -0.25 - 0.34$

$= -0.59\ V$

This negative value of E°_{cell} means this reaction is not feasible. So it is not possible for copper to reduce nickel(II) ions in solution to nickel.

23.3 The Nernst equation

The Nernst equation

The Nernst equation is used to predict quantitatively how the value of an electrode potential varies with the concentrations of the aqueous ions.

The Nernst equation is:

$$E = E_0 + \frac{0.059}{z} \log\left(\frac{[\text{oxidised species}]}{[\text{reduced species}]}\right)$$

where

E is the cell potential in volts under non standard conditions

E_0 is the cell potential in volts under standard conditions

R is the universal gas constant (i.e. 8.31 J K^{-1} mol^{-1})

z is the number of moles of electrons exchanged in the electrochemical reaction (in mol)

[oxidised species] is the concentration in mol dm^{-3} of the aqueous ions that are oxidised

[reduced species] is the concentration in mol dm^{-3} of the aqueous ions that are reduced

Examples

1 A cell consists of a zinc electrode in a solution of Zn^{2+} ions with concentration 0.80 mol dm^{-3} connected by a salt bridge to a solution Ag^{+} ions with concentration 1.30 mol dm^{-3} into which a silver electrode is immersed.

The standard electrode potentials for the electrodes are as follows:

$Zn^{2+}(aq) + 2e^{-} \longrightarrow Zn(s)$ $\quad E^{\circ} = -0.76\ V$

$Ag^{+}(aq) + e^{-} \longrightarrow Ag(s)$ $\quad E^{\circ} = +0.80\ V$

a) Write the balanced equation for the cell reaction.

b) Calculate E°_{cell}.

c) Determine the initial voltage produced by this cell.

Answer

1 a) As E° for the Ag electrode is higher than the E° for the zinc electrode, electrons will flow from the Zn to the Ag electrode. The zinc will lose electrons to become zinc ions (i.e. oxidation occurs) and the Ag^{+} ions will gain electrons to become silver (i.e. reduction occurs).

When writing the two equations we have to ensure that the electron numbers in both reactions are the same. Hence, we have

$$Zn(s) \longrightarrow Zn^{2+}(aq) + 2e^{-}$$

$$2Ag^{+}(aq) + 2e^{-} \longrightarrow 2Ag$$

Combining these two equations and cancelling the electrons on both sides we have the overall equation

$$Zn(s) + 2Ag^{+}(aq) \longrightarrow 2Ag + Zn^{2+}(aq)$$

b) The shorthand representation for the cell can be written as follows

$$Zn(s)\,|\,Zn^{2+}(aq)\,||\,Ag^{+}(aq)\,|\,Ag(s)$$

Then $E^{\ominus}_{cell} = E^{\ominus}_{\text{right-hand electrode}} - E^{\ominus}_{\text{left-hand electrode}}$

$= 0.80 - (-0.76)$

$= 1.56\text{V}$

c) Using the Nernst equation

$$E = E_o + \frac{0.059}{z}\log\left(\frac{[\text{oxidised species}]}{[\text{reduced species}]}\right)$$

we have

$$E = 1.56 + \frac{0.059}{2}\log\left(\frac{[0.80]}{[1.30]^2}\right)$$

Note that as there are 2 moles of electrons exchanged in the balanced equation, $z = 2$.

Also, from the balanced equation there is one mole of the oxidised species and two moles of the reduced species and this means that the concentration of the reduced species is raised to the power 2 (i.e. $[\text{reduced species}]^2$).

Hence $E = 1.56 + \frac{0.059}{2}\log\left(\frac{[0.80]}{[1.30]^2}\right)$

$= 1.56 - \frac{0.059}{2}\log 0.47$

$= 1.57\text{ V}$

» Pointer

Note that the electrode with the more positive potential is written on the right.

2 The diagram shows the apparatus used to measure the E_{cell} of a cell consisting of a standard Cu^{2+}/Cu electrode and an Ag^{+}/Ag electrode.

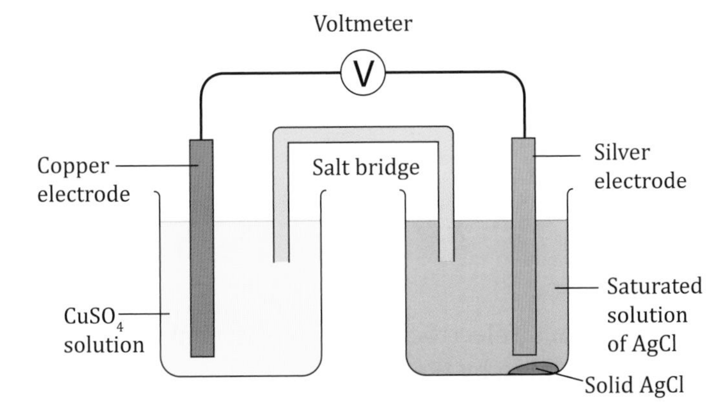

a) If the cell was operating under standard conditions, use the standard electrode potentials for the half-equations shown below to calculate the $E^{\ominus}_{cell}$.

$Cu^{2+}(aq) + 2e^{-} \longrightarrow Cu(s)$ $E^{\ominus} = +0.34\text{ V}$

$Ag^{+}(aq) + e^{-} \longrightarrow Ag(s)$ $E^{\ominus} = +0.80\text{ V}$

b) The Ag^+/Ag electrode is **not** under standard conditions owing to the $[Ag^+]$ in a standard solution of AgCl being much less than 1 mol dm^{-3}.

In the experiment shown in the diagram, E_{cell} was measured and found to be +0.17 V.

i) Calculate the value of $E_{electrode}$ for the Ag^+/Ag electrode.

ii) Use the equation $E_{electrode} = E^{\ominus}_{electrode} + 0.06 \log [Ag^+]$ to calculate $[Ag^+]$ in the saturated solution.

Answer

2 a) The standard electrode potential of silver (i.e. +0.80 V) is higher than that of copper (i.e. +0.34 V), so the electrons will flow from the copper to the silver.

The shorthand for the cell is

$Cu(s) \mid Cu^{2+}(aq) \mid\mid Ag^+(aq) \mid Ag(s)$

$$E^{\ominus}_{cell} = E^{\ominus}_{\text{right-hand electrode}} - E^{\ominus}_{\text{left-hand electrode}}$$

$$= +0.80 - 0.34$$

$$= 0.46 \text{ V}$$

b) i) The E_{cell} has been reduced by $0.46 - 0.17 = 0.29$ V and this is due to the lower electrode potential at the Ag electrode.

For the Ag electrode, $E_{electrode} = 0.80 - 0.29 = 0.51$ V

ii)

$$E_{electrode} = E^{\ominus}_{electrode} + 0.06 \log [Ag^+]$$

$$0.51 = 0.80 + 0.06 \log [Ag^+]$$

$$0.51 - 0.80 = 0.06 \log [Ag^+]$$

$$\frac{-0.29}{0.06} = \log [Ag^+]$$

$$-4.833 = \log [Ag^+]$$

To remove the log function press shift and then log and then enter the number −4.833. Note that this finds $10^{-4.833}$.

Hence $[Ag^+] = 1.47 \times 10^{-5}$ mol dm^{-3}

23.4 Electric charge and the Faraday constant

Electric charge

When a current flows in an electrochemical cell, the amount of charge, Q, in Coulombs that flows is given by the equation

$Q = It$ where I is the current in Amps and t is the time for which the current flows in seconds.

The Faraday constant

The Faraday constant, F, is the magnitude of electric charge per mole of electrons and has the value of 96 485 C mol^{-1}.

The Faraday constant, F, is related to the Avogadro number L (sometimes also called N_A) and the charge on an electron, e, by the formula

$$F = L\,e$$

During electrolysis the ions are 'forced' to undergo either oxidation (at the anode) or reduction (at the cathode). Using the Faraday constant, it is possible to calculate the amount in moles of a substance that has been oxidised.

Examples

1 a) By writing an equation, state the relationship between the Faraday constant, F, the Avogadro constant, L, and the charge on the electron, e

b) During an experiment, the Faraday constant was found to be the following value:

$$F = 9.63 \times 10^{4}\ \text{C mol}^{-1}.$$

Taking the charge on the electron, $e = 1.60 \times 10^{-19}$ C, calculate a value for the Avogadro constant, L.

Answer

1 a) $F = L\,e$

b) $L = \frac{F}{e} = \frac{9.63 \times 10^{4}}{1.6 \times 10^{-19}} = 6.02 \times 10^{23}\ \text{mol}^{-1}$

2 A 40.0 amp current flowed through molten iron(III) chloride for 10.0 hours (36,000 s) producing chlorine gas at the anode and iron at the cathode. Determine the mass of iron in grams that is produced during this time.

Answer

2 Writing the half-reactions that take place at the anode and at the cathode we have:

At the anode, oxidation occurs: $2Cl^{-} \longrightarrow Cl_2(g) + 2e^{-}$

At the cathode, reduction occurs: $Fe^{3+} + 3e^{-} \longrightarrow Fe(s)$

The charge in C flowing as a result of the current is given by Coulomb's law

$Q = I\,t$ where Q is the charge in Coulombs, I is the current in Amps and t is the time in seconds.

Hence $Q = I\,t = 40.0 \times 36000 = 1.44 \times 10^{6}$ C

Now one mole of electrons has a charge of the Faraday constant (i.e. 96485 C).

Hence, number of moles of electrons with a total charge of

$$1.44 \times 10^{6}\ \text{C} = \frac{1.44 \times 10^{6}}{96\,485} = 14.9$$

According to the half equation for the iron, three moles of electrons produce one mole of iron.

As we know the number of moles of electrons, we divide this by 3 to work out the number of moles of Fe.

Hence, number of moles of Fe $= \frac{14.9}{3} = 4.97$.

Using a periodic table, each mole of Fe has a mass of 55.85 g.

Hence, mass of Fe produced $= 4.97 \times 55.85 = 278$ g

Test yourself 23

1. The half-equations that occur in a lead-acid cell are shown here:

   $PbO_2(s) + 3H^+(aq) + HSO_4^-(aq) + 2e^- \longrightarrow PbSO_4(s) + 2H_2O(l)$ $\quad E^\ominus = 1.69\ V$

   $PbSO_4(s) + H^+(aq) + 2e^- \longrightarrow Pb(s) + HSO_4^-$ $\quad E^\ominus$ to be found

   If it is known that the $PbO_2/PbSO_4$ electrode is the positive terminal of the cell and that the emf of the cell is 2.15 V, calculate the value of the missing standard electrode potential for the half-equation shown above.

2. Use data in the table of electrode half-equations and their corresponding standard electrode potentials on page 136 for this question.
   (a) What is the emf produced by a cell consisting of an aluminium electrode in contact with a solution of Al^{3+} ions and an iron electrode in contact with a solution of Fe^{2+} ions?
   (b) State which of the electrodes will be the positive electrode.

3. Use data in the table of electrode half-equations and their corresponding standard electrode potentials on page 136 for this question.

   The cell represented below was set up:

   $Pt(s) \mid Fe^{2+}(aq), Fe^{3+}(aq) \parallel Cl_2(g), 2Cl^-(aq) \mid Pt(s)$

   Determine the emf of the above cell.

4. The table below shows some redox half-equations and standard electrode potentials:

Electrode half-equation	Standard electrode potential $E^\ominus$/V
$F_2(g) + 2e^- \longrightarrow 2F^-(aq)$	+2.87
$H_2O_2(aq) + 2H^+(aq) + 2e^- \longrightarrow 2H_2O(l)$	+1.77
$Cl_2(g) + 2e^- \longrightarrow 2Cl^-(aq)$	+1.36
$O_2(g) + 4H^+(aq) + 4e^- \longrightarrow 2H_2O(l)$	+1.23
$Fe^{2+}(aq) + 2e^- \longrightarrow Fe(s)$	−0.44
$Zn^{2+}(aq) + 2e^- \longrightarrow Zn(s)$	−0.76
$Al^{3+}(aq) + 3e^- \longrightarrow Al(s)$	−1.66

   (a) Using the data in the table identify which is the strongest reducing agent and explain how you arrived at your answer.
   (b) Fluorine reacts with water. Using the data in the table, explain why and write an equation for the reaction that occurs.

5. Using the data in the following table determine whether or not the following reaction will occur:

   $Fe^{3+}(aq) + Cl^-(aq) \longrightarrow Fe^{2+}(aq) + \frac{1}{2}Cl_2(g)$

$Cl_2(g) + 2e^- \longrightarrow 2Cl^-(aq)$	$E^\ominus = +1.36\ V$
$Fe^{3+}(aq) + e^- \longrightarrow Fe^{2+}(aq)$	$E^\ominus = +0.77\ V$

Chapter 24 Entropy and free energy change

24.1 Entropy

Entropy is a measure of the degree of disorder of a system. Systems become more energetically stable when they become more disordered, so for example, a gas has more entropy than a liquid, which has more entropy than a solid.

When a solid dissolves in a liquid to form ions, the ions in solution have more entropy as they are more disordered. If there is a reaction where there are a greater number of total moles of products compared with the reactants, the entropy will have increased.

Two factors increase entropy:

- An increased number of particles.
- Increasing the temperature.

A perfectly ordered crystal at absolute zero has zero entropy so it is possible to find an absolute entropy. Tables of absolute entropy values S are available. It is important to note that entropy is measured in $J\,K^{-1}mol^{-1}$.

Entropy change

The entropy change for a reaction is the difference between the entropy of the products and the entropy of the reactants.

Hence, $\Delta S = S_{products} - S_{reactants}$

When the entropy is determined under standard conditions, it is called the standard entropy, $S^{\ominus}$ and the above equation can be written as

$$\Delta S^{\ominus} = S^{\ominus}_{products} - S^{\ominus}_{reactants}$$

Pointer

ΔS is the change in entropy and the S values are absolute entropy values.

Examples of entropy changes

When water boils to form steam the entropy increases, so ΔS is positive.

When water changes to ice at 0 °C, the entropy decreases, so ΔS is negative.

In the reaction for the production of ammonia:

$$N_2(g) + 3H_2(g) \rightleftharpoons 2NH_3(g)$$

There are fewer particles on the right of the equation compared with the left, so the disorder has decreased, hence ΔS is negative.

24.2 Gibbs free energy

The Gibbs free energy, ΔG, measured in J mol^{-1} can be used to determine whether a reaction is spontaneous or not. If ΔG is negative then the reaction is spontaneous. If it is positive, the reaction will not occur. If $\Delta G = 0$, then the reaction will just be on the point of occurring.

The Gibbs free energy ΔG is related to the enthalpy ΔH, the entropy ΔS and the absolute temperature T by the equation:

$$\Delta G = \Delta H - T\Delta S$$

Examples

1 When heated Al_2Cl_6(g) molecules dissociate into $AlCl_3$(g) molecules:

$$Al_2Cl_6(g) \rightleftharpoons 2AlCl_3(g)$$

a) Explain why the entropy of this gaseous system is increasing.

b) By using the equation $\Delta G = \Delta H - T\Delta S$, calculate the temperature at which the dissociation of Al_2Cl_6(g) into $2AlCl_3$(g), just occurs.

The entropy change for the above reaction, ΔS is 88 J mol^{-1} K^{-1} and the enthalpy change, ΔH is 60 kJ mol^{-1}.

Answer

1 a) The number of molecules increases from 1 on the left-hand side to 2 on the right-hand side of the equation. Greater numbers of molecules increase disorder so the entropy increases.

b) $\Delta G = \Delta H - T\Delta S$

On the point of the reaction just occurring, $\Delta G = 0$.

So, $0 = 60000 - T \times 88$

Hence, $T = 682$ K (3 s.f.)

» *Pointer*
Always be careful with values of ΔH in kJ mol^{-1}. Remember to multiply the value by 1000 to convert the value to units in J mol^{-1}.

2 The following graph shows how the free energy for a reaction changes with temperature above 140 K:

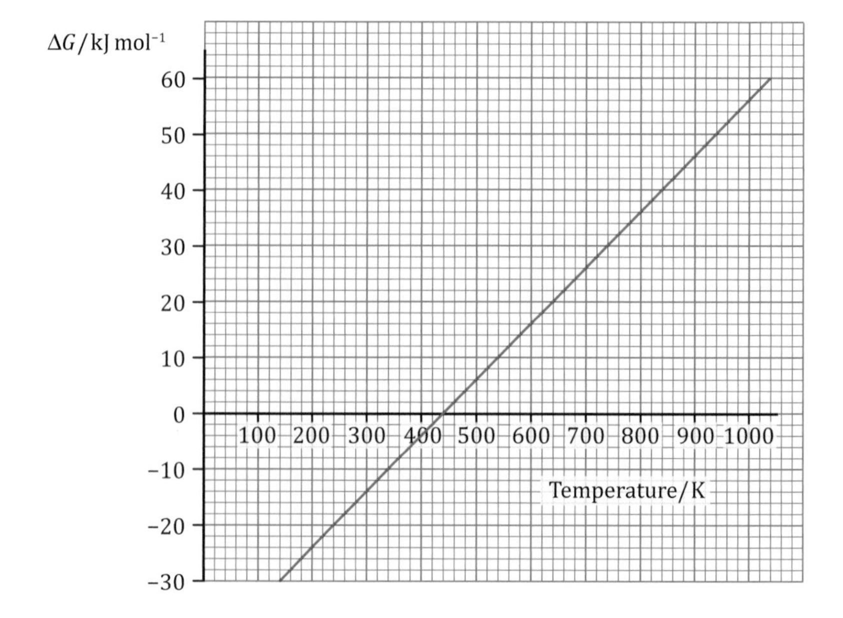

Write the equation which connects the quantities ΔG, ΔH and ΔS.

Find the gradient of the graph and give the name and units of the quantity that this gradient represents.

Answers

2 a) $\Delta G = \Delta H - T\Delta S$

b) Gradient of graph $= \dfrac{20}{200} = 0.1$

Units will be $\dfrac{\text{kJ mol}^{-1}}{\text{K}} = \text{kJ mol}^{-1}\,\text{K}^{-1}$

The quantity is the enthalpy change $-\Delta S$

If you are unsure about working out the gradient, then you need to look at page 39 of the maths section. All straight line graphs have an equation of the form $y = mx + c$. Rearranging the equation $\Delta G = \Delta H - T\Delta S$ into this format gives $\Delta G = -\Delta S\, T + \Delta H$, so the gradient is $-\Delta S$.

The units will be the units of the quantity on the y-axis divided by the units of the quantity on the x-axis.

Test yourself 24

1. Here is a table containing thermodynamic data for the reaction between hydrogen and oxygen to form steam:

	$S^{\ominus}$/J K^{-1} mol^{-1}	$\Delta H_f^{\ominus}$/kJ mol^{-1}
$H_2(g)$	131	0
$O_2(g)$	205	0
$H_2O(g)$	189	−242

Here is the equation for the formation of one mole of steam:

$H_2(g) + \frac{1}{2}O_2(g) \longrightarrow H_2O(g)$

There is a temperature beyond which the above reaction is not feasible. Calculate this temperature giving your answer in Kelvin.

2. Ammonia reacts with oxygen according to the following equation:

$4NH_3(g) + 5O_2(g) \longrightarrow 4NO(g) + 6H_2O(g)$

The standard entropy values are

$S^{\ominus}(NH_3) = 193\ \text{J K}^{-1}\text{mol}^{-1}$

$S^{\ominus}(O_2) = 205\ \text{J K}^{-1}\text{mol}^{-1}$

$S^{\ominus}(NO) = 211\ \text{J K}^{-1}\text{mol}^{-1}$

$S^{\ominus}(H_2O) = 189\ \text{J K}^{-1}\text{mol}^{-1}$

(a) Calculate the standard entropy of the reactants.

(b) Calculate the standard entropy of the products.

(c) Calculate the standard entropy change for the reaction.

3. Methanol undergoes the following reaction with oxygen.

$CH_3OH(l) + \frac{3}{2}O_2(g) \longrightarrow CO_2(g) + 2H_2O(l)$

The standard enthalpy change of combustion for the above reaction is −727 kJ mol^{-1}.

The standard entropy change for the above reaction is −81 J K^{-1} mol^{-1}.

(a) Use the above information to calculate a value for the Gibbs free energy ΔG for this reaction.

(b) Explain the significance of the sign of ΔG for this reaction.

Chapter 25

Equilibria

25.1 Reactions in equilibrium

Many reactions involve a forward reaction and a backward reaction. When the rates of the forward reaction and the backward reaction are equal, the reaction is said to be in equilibrium and the concentrations of the reactants and products remain constant.

A reversible reaction involving two reactants and two products can be written in the following way:

$a\,A + b\,B \rightleftharpoons c\,C + d\,D$ where a, b, c and d are the numbers of moles of A, B, C and D respectively.

The following equation connects the equilibrium constant, K_c, with the concentrations of A, B, C and D when the stoichiometry (i.e. the letters a, b, c and d in the example above) of the equation is known:

» Pointer
[A] means the concentration of A.
[B] means the concentration of B, etc. Concentrations are measured in the units $mol\,dm^{-3}$.

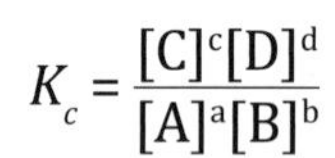

$$K_c = \frac{[C]^c[D]^d}{[A]^a[B]^b}$$

Calculating the units of the equilibrium constant K_c

The units of the equilibrium constant K_c depend on the stoichiometry of the equation (i.e. the values of the numbers a, b, c and d).

Take the following reversible reaction:

$$H_2(g) + I_2(g) \rightleftharpoons 2HI(g)$$

By comparing the stoichiometry of this reaction with:

$$aA + bB \rightleftharpoons cC + dD$$

you can see that $a = 1$, $b = 1$ and $c = 2$ (note there is only one product so there is no D or d).

Putting these values into the expression for the equilibrium constant, we obtain:

$$K_c = \frac{[C]^2}{[A][B]}$$

Now concentrations have the units of $mol\,dm^{-3}$ and the units without the values can be entered into the expression for the equilibrium constant K_c.

» Pointer
The units on the top cancel with the units on the bottom, so there are no units for the equilibrium constant for this reaction.

$$K_c = \frac{[mol\,dm^{-3}]^2}{[mol\,dm^{-3}][mol\,dm^{-3}]} = \frac{[mol\,dm^{-3}]^2}{[mol\,dm^{-3}]^2} = \text{no units}$$

Take the following equilibrium, for example:

$$2SO_2(g) + O_2(g) \rightleftharpoons 2SO_3(g)$$

The expression for the equilibrium constant for this reaction is:

$$K_c = \frac{[SO_3]^2}{[SO_2]^2[O_2]}$$

A quick look at this expression will reveal that the units for the two concentrations that are squared will cancel with each other because the units will be the same.

Only the $[O_2]$ in the denominator will have units.

Hence, we have units of $K_c = \frac{1}{[O_2]} = \frac{1}{\text{mol dm}^{-3}} = \text{mol}^{-1}\text{ dm}^3$

» Pointer

Notice the way the units on the bottom of the fraction (i.e. mol dm^{-3}) have their powers changed when brought to the top (i.e. they change to mol^{-1}dm^3)

Examples

1 In the Haber process for the industrial production of ammonia the following equilibrium is set up by reacting nitrogen and hydrogen gases over an iron catalyst:

$N_2(g) + 3H_2(g) \rightleftharpoons 2NH_3(g)$

a) Write the expression for K_c for the above reaction.

b) Determine the units of K_c.

Answer

1 a) $K_c = \frac{[NH_3]^2}{[N_2][H_2]^3}$

b) Units for $K_c = \frac{[\text{mol dm}^{-3}]^2}{[\text{mol dm}^{-3}][\text{mol dm}^{-3}]^3}$

$= \frac{[\text{mol dm}^{-3}]^2}{[\text{mol dm}^{-3}]^4}$

$= \frac{1}{[\text{mol dm}^{-3}]^2}$

$= \frac{1}{\text{mol}^2\text{ dm}^{-6}}$

$= \text{mol}^{-2}\text{ dm}^6$

» Pointer

When the units are brought from the bottom to the top the powers of the units must have their signs reversed.

2 A mixture of gaseous hydrogen, iodine and hydrogen iodide was left to reach equilibrium in a sealed container of volume V dm^3. The mixture was left to reach equilibrium at a certain temperature and the equilibrium mixture contained 0.19 mol of iodine, 0.38 mol of hydrogen and 1.94 mol of hydrogen iodide. The equation for the reaction is as follows:

$H_2(g) + I_2(g) \rightleftharpoons 2HI(g)$

a) Write down an expression for the equilibrium constant K_c for the above equilibrium.

b) Calculate the value of the equilibrium constant K_c at this temperature.

Answer

2 a) $H_2(g) + I_2(g) \rightleftharpoons 2HI(g)$

$K_c = \frac{[HI]^2}{[I_2][H_2]}$

» Pointer

Note the way we use the number of moles as the concentrations here. We can do this because the volume is fixed as the container is sealed.

b) $K_c = \frac{[HI]^2}{[I_2][H_2]} = \frac{[1.94]^2}{[0.19][0.38]} = 52.1$

3 The following equilibrium reaction was established in a sealed container at temperature *T*:

$$A(g) + 2B(g) \rightleftharpoons 2C(g)$$

The equilibrium mixture contained 4.15 mol of A and 6.09 mol of C and the equilibrium constant K_c for the reaction was 65.7 mol^{-1} dm^3.

a) Write down an expression for the equilibrium constant K_c for the above equilibrium.

b) If the volume of the container was 10.0 dm^3, calculate the concentration in mol dm^{-3} of B in the equilibrium mixture.

c) Deduce the value of the equilibrium constant at temperature *T* for the following reaction.

$$2C(g) \rightleftharpoons A(g) + 2B(g)$$

Pointer
Note that there are no units for the equilibrium constant for this reaction as the concentration units on the top and bottom of the fraction cancel.

Answer

3 a) $K_c = \frac{[C]^2}{[A][B]^2}$

b) Rearranging the above equation for $[B]^2$ gives

$$[B]^2 = \frac{[C]^2}{[A]K_c} = \frac{\left[\frac{6.09}{10.0}\right]^2}{\left[\frac{4.15}{10.0}\right] \times 65.7} = 0.0136$$

$$[B] = \sqrt{0.0136} = 0.117 \text{ mol dm}^{-3}$$

c) Equilibrium constant $= \frac{[A][B]^2}{[C]^2}$

Note that the concentrations of reactants and products have swapped places compared to the expression in part (a).

Hence the equilibrium constant $= \frac{1}{K_c} = \frac{1}{65.7} = 0.01522$

Pointer
Note that the gases are in a sealed container which will have a fixed volume. Hence the number of moles can be used as the concentrations in the equation for K_c.

Pointer
Notice that we use the formula to work out the concentrations of C and A (i.e. [A] and [C]).

Pointer
Note that in the question it asks for the value of the equilibrium constant. This means we are not expected to include the units in the answer. The units would have been the opposite for the units for K_c in part (a). They would be mol dm^{-3}.

25.2 Finding the concentrations at equilibrium

In many questions you will be given information about the initial amounts of reactants and products and then use these to find the concentrations after equilibrium has been reached. The following example shows how equilibrium concentrations can be found.

Example

The ester ethyl propanoate was prepared by mixing 1.20 mol of propanoic acid, 2.50 mol of ethanol and 4.00 mol of water. At a fixed temperature, the mixture was left to reach equilibrium according to the following equation:

$$CH_3CH_2COOH + CH_3CH_2OH \rightleftharpoons CH_3CH_2COOCH_2CH_3 + H_2O$$

At equilibrium the mixture contained 0.45 mol of the ester ethyl propanoate.

a) Calculate the number of moles of each of the following in the equilibrium mixture:

i) Propanoic acid

ii) Ethanol

iii) Water

b) Write an expression for the equilibrium constant, K_c, for this reaction.

c) Calculate the equilibrium constant at this temperature and explain why it has no units.

Answer

1 a) The 4.00 mol of water are not needed as it is not part of the reactants.

This water will simply add to the number of moles produced at equilibrium.

First we write down the number of moles at equilibrium. Now 0.45 moles of ethyl propanoate are produced which means there will also be 0.45 moles of water produced which needs to be added to the 4.00 moles initially present. This gives 0.45 + 4.00 = 4.45 moles of water.

Now the 0.45 moles of each product means that the number of moles of each reactant needs to be reduced by this amount. Hence the number of moles of acid at equilibrium is 1.20 – 0.45 = 0.75 moles and the number of moles of alcohol is 2.50 – 0.45 = 2.05 moles.

Usually we work this out under the equation like this:

	CH_3CH_2COOH	+ $CH_3CH_2OH \rightleftharpoons$	$CH_3CH_2COOCH_2CH_3$	+ H_2O
At equilibrium	1.20 – 0.45	2.50 – 0.45	0.45	0.45 + 4.00
	= 0.75 moles	= 2.05 moles	= 0.45 moles	= 4.45 moles

i) 0.75 moles

ii) 2.05 moles

iii) 4.45 moles

b) $K_c = \dfrac{[CH_3CH_2COOCH_2CH_3][H_2O]}{[CH_3CH_2COOH][CH_3CH_2OH]}$

c) $K_c = \dfrac{[CH_3CH_2COOCH_2CH_3][H_2O]}{[CH_3CH_2COOH][CH_3CH_2OH]}$

$= \dfrac{[0.45][4.45]}{[0.75][2.05]}$

$= 1.30$ (3 s.f.)

If you look at the expression for the equilibrium constant

$K_c = \dfrac{[CH_3CH_2COOCH_2CH_3][H_2O]}{[CH_3CH_2COOH][CH_3CH_2OH]}$

The units for concentration will be squared on the top and as they are also squared on the bottom, the units will cancel.

Hence K_c in this case has no units.

25.3 Using partial pressures rather than concentrations

For gaseous reactions, it is more convenient to replace the concentrations with the partial pressures of the gases. When partial pressures are used rather than concentrations the equilibrium constant is called K_p.

In a mixture of three gases A, B and C then the partial pressure of gas A is the pressure gas A would exert if it occupied the entire volume on its own.

By adding the partial pressures of the gases, the total pressure, P, of the mixture of gases is obtained.

Hence we have:

$$P = p_A + p_B + p_C$$

Suppose we have the following equilibrium we met earlier in the chapter:

$$H_2(g) + I_2(g) \rightleftharpoons 2HI(g)$$

The equilibrium constant, K_p, can be calculated by using the partial pressures rather than the concentrations. Hence we have

$$K_p = \frac{p(HI(g))^2}{p(H_2(g)) \times p(I_2(g))}$$

Example

The gas nitrogen dioxide (NO_2) exists in equilibrium with dinitrogen tetroxide (N_2O_4).

The equation for this is:

$$2NO_2(g) \rightleftharpoons N_2O_4(g)$$

a) Write an equation for the equilibrium constant, K_p, for this equilibrium.

b) When the temperature is 373 K, the partial pressure of pure NO_2 was 6.00×10^5 Pa. When in dynamic equilibrium, the partial pressure of the NO_2 that remained was 5.62×10^5 Pa. Calculate the value of K_p giving its units.

Answer

a) $K_p = \frac{pN_2O_4}{pNO_2{}^2}$

b) From the stoichiometry of the equation, you can see that 2 moles of NO_2 are in equilibrium with 1 mole of N_2O_4.

Now as the partial press of NO_2 at equilibrium is 5.62×10^5 Pa, there is a difference in pressure of $6.00 \times 10^5 - 5.62 \times 10^5 = 0.38 \times 10^5$ Pa.

Because there are twice the number of moles of NO_2 compared to N_2O_4, the partial pressure of N_2O_4 will be half of 0.38×10^5 Pa. Partial pressure of N_2O_4 is 0.19×10^5 Pa.

$$K_p = \frac{pN_2O_4}{pNO_2{}^2} = \frac{0.19 \times 10^5}{(5.62 \times 10^5)^2} = 6.02 \times 10^{-8}$$

$$K_p = \frac{pN_2O_4}{pNO_2{}^2} = \frac{Pa}{(Pa)^2} = \frac{1}{Pa} = Pa^{-1}$$

» *Pointer*

The units for the partial pressure are entered into the formula for K_p to determine the units of K_p.

Test yourself 25

1. When phosphorus(V) chloride is heated it thermally dissociates partially to form phosphorus(III) chloride and chlorine.

   The reaction reaches equilibrium and is represented by the following reaction:

   $PCl_5(g) \rightleftharpoons PCl_3(g) + Cl_2(g)$

   (a) Write down an expression for the equilibrium constant K_c for the above equilibrium.

   (b) Calculate the units of the equilibrium constant K_c.

2. At a temperature T, the reaction

   $C_2H_5OH(l) + C_2H_5COOH(l) \rightleftharpoons C_2H_5COOC_2H_5(l) + H_2O(l)$

   is in dynamic equilibrium.

   At the start of the above reaction, 4.0 moles of ethanol were allowed to react with 1.0 mole of ethanoic acid. When dynamic equilibrium was established, the amount of ethanoic acid remaining was 0.07 moles. Calculate the equilibrium constant for this reaction at temperature T.

3. Dinitrogen tetroxide (N_2O_4) decomposes on heating to form nitrogen dioxide (NO_2). The reaction takes place in a sealed reaction vessel of volume 1 dm^3. Dynamic equilibrium is reached as shown by the following equation:

   $N_2O_4(g) \rightleftharpoons 2NO(g)$

   (a) Write the expression for the equilibrium constant K_c.

   (b) Determine the units of K_c.

   (c) If when dynamic equilibrium is reached, there are 6 moles of both gases, calculate the value of K_c.

4. Ammonia is manufactured from nitrogen and hydrogen using the Haber process according to the following:

   $N_2(g) + 3H_2(g) \rightleftharpoons 2NH_3(g)$

   (a) Determine the equilibrium expression for this reaction.

   (b) Determine the units for the equilibrium constant K_c.

   (c) The above reaction took place in a sealed reaction vessel of volume 6 dm^3 at a temperature of 500 K and a high pressure. The reaction was left to reach equilibrium and at equilibrium there were 12.0 mol of H_2 and 7.2 mol of N_2.

   At the temperature of the reaction (500 K), the value of the equilibrium constant was 0.08. Calculate the amount in mol of ammonia in the equilibrium mixture at this temperature.

Chapter 26 Calculations involving pH

26.1 pH and the pH scale

Pointer

The concentration of the hydrogen (i.e. H^+) ions is enclosed in the square brackets.

Pointer

The $\log_{10}$is marked log on most calculators. To find the pH press the key marked (–) and then the key marked log and then enter the value for the concentration followed by equals. Try this out using the concentration 0.001.

$-\log_{10}[0.001]$

Check using your calculator that you obtain the value 3 for the pH.

Pointer

You can see from the table that an increase in 1 in the pH scale corresponds with a ten-fold decrease in the hydrogen ion concentration.

Pointer

Remember that the more negative the power of 10 for a number in standard form, the smaller the number.

The definition of pH:

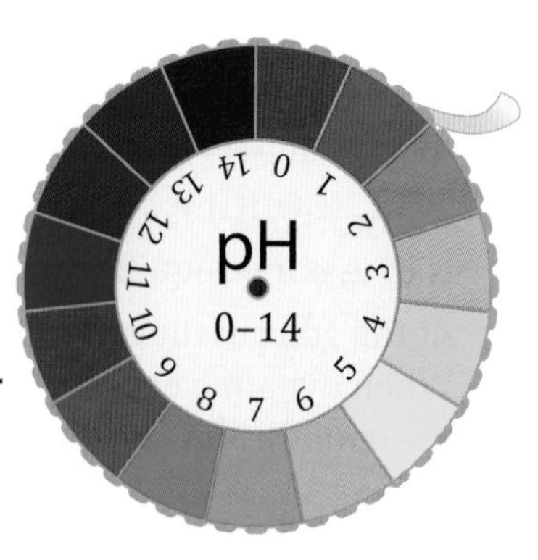

pH is the minus logarithm to the base 10 of the hydrogen ion concentration.

Hence, pH = $-\log_{10}[H^+]$

The pH scale runs from 0 to 14. On the pH scale values below 7 represent acidic solutions, a value equal to 7 equals a neutral solution and values over 7 represent alkaline solutions.

The following table shows some hydrogen ion concentrations along with the corresponding pH:

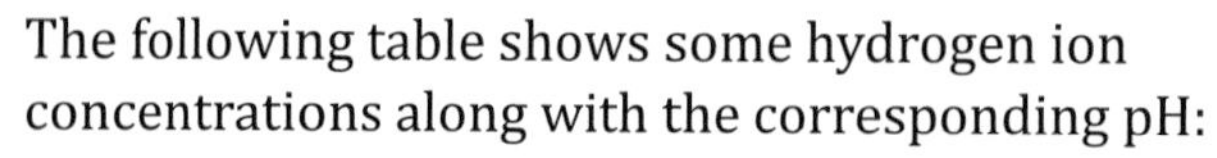

$[H^+]$ in mol dm^{-3}	pH
1	0
1×10^{-1}	1
1×10^{-5}	5
1×10^{-6}	6
1×10^{-7}	7
1×10^{-8}	8
1×10^{-14}	14

Looking at the above table, you can see:

A low hydrogen ion concentration results in a high pH.

A high hydrogen ion concentration results in a low pH.

Working out pH from the hydrogen ion concentration

If an acid is a strong acid such as HNO_3(aq), HCl(aq) or H_2SO_4(aq) the acid is completely ionised in solution. So, for example, in H_2SO_4(aq) the following ions would be produced:

Pointer

This is not an equilibrium. As the acid is strong it is completely dissociated in solution.

$$H_2SO_4(aq) \longrightarrow 2H^+(aq) + SO_4^{2-}(aq)$$

The pH can be calculated from the hydrogen ion concentration using the formula:

$$pH = -\log_{10}[H^+]$$

It is important to note that in the above reaction the number of moles of H^+ ions will be double the number of moles of H_2SO_4 owing to the stoichiometry of the equation.

It is also important to note that the concentration of the hydrogen ions (written as $[H^+]$) must be in the units of mol dm^{-3}.

Examples

1 A strong acid HX in aqueous solution has a concentration of 0.185 mol dm^{-3} at a certain temperature. Calculate the pH of this solution giving your answer to 2 decimal places.

Pointer

From the equation, you can see that a certain number of moles of HX will produce the same number of moles of H^+ ions.

Answer

1 The equation for the formation of H^+ ions is as follows

$HX(aq) \longrightarrow H^+(aq) + X^-(aq)$

In a solution of 0.185 mol dm^{-3} of HX there is a concentration of 0.185 mol dm^{-3} of H^+ ions.

$pH = -\log_{10}[H^+] = -\log_{10}[0.185] = 0.73$ (2 d.p.)

Pointer

Always check to see whether the answer should be given to a certain number of decimal places or significant figures. If you are not told this in the question, then most final answers should be given to three significant figures.

2 Calculate the pH of the solution formed when 35.0 cm^3 of 0.185 mol dm^{-3} aqueous nitric acid are added to 25.0 cm^3 of 0.150 mol dm^{-3} aqueous potassium hydroxide at 25 °C. Assume that the nitric acid is fully dissociated.

Answer

2 You first need to work out the number of moles of nitric acid and the number of moles of potassium hydroxide.

Number of moles of nitric acid in 35.0 cm^3, $n = \frac{V}{1000} \times c$

$= \frac{35.0}{1000} \times 0.185$

$= 6.475 \times 10^{-3}$ mol

Pointer

In this formula, the volume, V is in cm^3.

Number of moles of potassium hydroxide in 25.0 cm^3, $n = \frac{V}{1000} \times c$

$= \frac{25.0}{1000} \times 0.150$

$= 3.75 \times 10^{-3}$ mol

Pointer

This question is about a neutralisation reaction. You can see that both the concentrations and volumes of the acid and alkali are given so it is possible to determine the number of moles of each. You can then see if either the acid or alkali is in excess.

The equation for the reaction is:

$HNO_3(aq) + KOH(aq) \longrightarrow KNO_3(aq) + H_2O(l)$

You can see that this is a neutralisation reaction. According to the equation, one mole of acid reacts with one mole of alkali.

If the number of moles of each is not the same then one of the reactants will be in excess, so we need to find which is in excess and by how many moles.

The nitric acid is in excess by $6.475 \times 10^{-3} - 3.75 \times 10^{-3} = 2.725 \times 10^{-3}$ mol

Pointer

The KOH contains the lesser number of moles. So, 3.75×10^{-3} mol of the alkali will react with, 3.75×10^{-3} mol of the acid. As the acid is in excess, the resulting solution will be acidic.

This number of moles of acid is in (25.0 + 35.0) cm^3 = 60 cm^3 = 0.06 dm^3 of solution.

Pointer

Remember to add both volumes together as the excess acid is dissolved in this volume.

Hence, concentration of acid $\frac{n}{V} = \frac{2.725 \times 10^{-3}}{0.06} = 4.542 \times 10^{-2}$ mol dm^{-3}

$pH = -\log_{10}[H^+] = -\log_{10}[4.542 \times 10^{-2}] = 1.34$ (3 s.f.)

3 A 30.0 cm^3 sample of 0.0550 mol dm^{-3} hydrochloric acid was placed in a beaker and distilled water was added until the pH of the solution was 1.75.

Calculate the total volume of the solution formed stating the units.

Answer

3 $pH = -\log_{10}[H^+]$ gives

$1.75 = -\log_{10}[H^+]$

$-1.75 = \log_{10}[H^+]$

Hence $10^{-1.75} = [H^+]$

$[H^+] = 0.0178 \text{ mol dm}^{-3}$

» Pointer
First we find the concentration of the acid which would give a pH of 1.75.

Number of moles of HCl in 30.0 cm³, $n = \frac{v}{1000} \times c$

$= \frac{30.0}{1000} \times 0.0550$

$= 1.65 \times 10^{-3}$ mol

As the acid is a strong acid, it is completely dissociated and [HCl] = $[H^+]$

Hence, [HCl] = 0.0178 mol dm^{-3}

» Pointer
The volume in this equation must be in cm³.

Hence $[HCl] = \frac{1000n}{v}$

$0.0178 = \frac{1000 \times 1.65 \times 10^{-3}}{v}$

So, $v = \frac{1000 \times 1.65 \times 10^{-3}}{0.0178}$

$= 92.7$ cm³ (3 s.f.)

Working out hydrogen ion concentration [H⁺] from the pH

Suppose we have an aqueous solution with a pH of 3 and we need to find the hydrogen ion concentration in mol dm^{-3}.

» Pointer
Multiply both sides by −1 so that the minus sign now appears on the left.

Substituting this value into $pH = -\log_{10}[H^+]$ gives

$3 = -\log_{10}[H^+]$

$-3 = \log_{10}[H^+]$

Now If $p = \log_b X$, then by definition $X = b^p$.

Hence $10^{-3} = [H^+]$

$[H^+] = 1 \times 10^{-3}$ mol dm^{-3}

» Pointer
Note that 1×10^{-3} can also be written as 0.001.

» Pointer
Use the calculator to work out 10^{-3}. On most calculators you will use the 10^x button which is usually obtained by pressing shift and then the log button. You then enter the (−) sign and then enter the number 3. Do this to check you get the answer.

The ionic product of water

Water can behave as both an acid (donating protons) or as a base (accepting protons). Water exists in the following equilibrium between undissociated water molecules (i.e. $H_2O(l)$) and acidic and basic ions (i.e. $H^+(aq)$ and $OH^-(aq)$).

$$H_2O(l) \rightleftharpoons H^+(aq) + OH^-(aq)$$

Water is only very slightly dissociated, so the concentrations of the ions are very low.

The equilibrium constant for the above equilibrium, K_c is given by:

$$K_c = \frac{[H^+(aq)][OH^-(aq)]}{[H_2O(l)]}$$

Now because only a small amount of water ionises, it can be assumed that the concentration of the water is constant and the value can be incorporated into a new constant, K_w, called the ionic product of water.

Hence we have:

$$K_w = [H^+(aq)][OH^-(aq)]$$

At 25 °C (298 K), K_w has the value 1.0×10^{-14} mol^2 dm^{-6}.

Hence, $[H^+(aq)][OH^-(aq)] = 1.0 \times 10^{-14}$ mol^2 dm^{-6}

In pure water, $[H^+(aq)] = [OH^-(aq)]$,

So we can write $[H^+(aq)]^2 = 1.0 \times 10^{-14}$

Hence, $[H^+(aq)] = 1.0 \times 10^{-7}$

and $[OH^-(aq)]^2 = 1.0 \times 10^{-14}$, thus giving $[OH^-(aq)] = 1.0 \times 10^{-7}$

» Pointer
Water dissociates to give equal numbers of moles and hence concentrations of H^+ and OH^- ions.

» Pointer
To remove a square you take the square root of both sides.

Finding the pH of alkaline solutions

To find the pH of an alkaline solution, you first need to find $[OH^-(aq)]$ (i.e. the concentration of the hydroxide ions in mol dm^{-3}). In the case of a strong alkali such as sodium hydroxide, the alkali is completely dissociated into ions.

$$NaOH(aq) \longrightarrow Na^+(aq) + OH^-(aq)$$

The number of moles of NaOH will equal the number of moles of OH^- ions so the concentrations will be the same.

You then need to use:

$$[OH^-(aq)][H^+(aq)] = 1.00 \times 10^{-14} \text{ mol}^2 \text{ dm}^{-6} \text{ to calculate } [H^+(aq)]$$

Then the equation $pH = -\log_{10}[H^+]$ is used to work out the pH.

» Pointer
This equation is the equation for the ionic product of water.

Example

5 g of NaOH pellets were dissolved in water and the solution made up with water to a total volume of 750 cm^3.

a) i) Calculate the number of moles in 5 g of NaOH.

ii) Determine the concentration of the solution in mol dm^{-3}.

b) i) Calculate the concentration of H^+ ions in the solution.

ii) Calculate the pH of the solution.

Answer

a) i) Number of moles, $n = \frac{m}{M} = \frac{5}{40.0} = 0.125$ mol

ii) Concentration, $c = \frac{1000n}{v} = \frac{1000 \times 0.125}{750} = 0.16666$

$= 0.167$ mol dm^{-3} (3 s.f.)

b) i) As NaOH(aq) is a strong alkali, it is completely dissociated.

The equation for this is:

» Pointer
The molar mass (or M_r) is worked out from the A_r of the component atoms looked up in the periodic table.

» Pointer
Notice that from the stoichiometry of the equation, the number of moles of NaOH and OH^- ions is the same.

Pointer

Divide both sides by 0.167 so that $[H^+(aq)]$ is made the subject of the equation.

$NaOH(aq) \longrightarrow Na^+(aq) + OH^-(aq)$

The concentration of the NaOH = concentration of OH^- ions

Hence, $[OH^-(aq)] = 0.167\ mol\ dm^{-3}$

ii) Now $[OH^-(aq)][H^+(aq)] = 1.00 \times 10^{-14}\ mol^2\ dm^{-6}$

Hence, $[0.167][H^+(aq)] = 1.00 \times 10^{-14}$

So, $[H^+(aq)] = \frac{1.00 \times 10^{-14}}{0.167} = 5.99 \times 10^{-14}\ mol\ dm^{-3}$

$pH = -\log_{10}[H^+] = -\log_{10}[5.99 \times 10^{-14}] = 13.2$ (3 s.f.)

Finding the pH of weak acids

Pointer

The stronger the acid is, the higher the value of K_c. We can tell this by looking at the fraction on the right-hand side. The numerator (top) will be larger as the concentrations of the ions will be larger and the denominator (bottom) will be smaller as more of the acid is dissociated.

Pointer

The units for K_a are always $mol\ dm^{-3}$.

Strong acids are completely dissociated whereas weak acids are not. Looking at the equation for a strong acid, the equation moves completely to the right whereas for a weak acid there exists an equilibrium between the acid and its ions. This equilibrium can be written in the following way:

$$HA(aq) \rightleftharpoons H^+(aq) + A^-(aq)$$

The equilibrium constant, K_c, for the above equilibrium can be written as:

$$K_c = \frac{[H^+(aq)]\ [A^-(aq)]}{[HA(aq)]}$$

As this dissociation refers to the dissociation of an acid, we replace K_c by K_a, called the acid dissociation constant.

Hence we can write $K_a = \frac{[H^+(aq)]\ [A^-(aq)]}{[HA(aq)]}$

Examples

1 Write an expression for the acid dissociation constant, K_a, for ethanoic acid $CH_3COOH(aq)$.

Answer

1 First write the equation for the equilibrium:

$CH_3COOH(aq) \rightleftharpoons H^+(aq) + CH_3COO^-(aq)$

$K_a = \frac{[H^+(aq)][CH_3COO^-(aq)]}{[CH_3COOH(aq)]}$

2 The acid dissociation constant, K_a, for methanoic acid (HCOOH) is $1.78 \times 10^{-4}\ mol\ dm^{-3}$ at 25 °C. Calculate the pH of a $0.0455\ mol\ dm^{-3}$ solution of methanoic acid.

Answer

2 First write an equation for the equilibrium:

$HCOOH(aq) \rightleftharpoons H^+(aq) + HCOO^-(aq)$

Then write the expression for the acid dissociation constant, K_a.

$K_a = \frac{[H^+(aq)][HCOO^-(aq)]}{[HCOOH(aq)]}$

Now $[H^+(aq)] = [HCOO^-(aq)]$

so $[H^+(aq)] = [HCOO^-(aq)]$ can be replaced by $[H^+(aq)]^2$

Hence, $K_a = \frac{[H^+(aq)]^2}{[HCOOH(aq)]}$

Rearranging to make $[H^+(aq)]^2$ the subject of the equation we obtain

$[H^+(aq)]^2 = K_a \times [HCOOH(aq)]$

$[H^+(aq)]^2 = 1.78 \times 10^{-4} \times 0.0455$

$[H^+(aq)] = \sqrt{1.78 \times 10^{-4} \times 0.0455}$

$= 2.85 \times 10^{-3}\ mol\ dm^{-3}$

$pH = -log_{10}[H^+] = -log_{10}[2.85 \times 10^{-3}] = 2.55$ (3 s.f.)

pK_a values

For weak acids we refer to a value called pK_a rather than pH because it is easier to determine.

To determine pK_a we use:

$$pK_a = -log_{10}K_a$$

This equation can be used to determine pK_a if K_a is known. The equation can also be rearranged and used in reverse to determine K_a if the pK_a is known and the equation for this is:

$$K_a = 10^{-pK_a}$$

Examples

1 Ethanoic acid, a weak acid, has a K_a value of $1.7 \times 10^{-5}\ mol\ dm^{-3}$. Calculate the p$K_a$ of this acid.

Answer

1 $pK_a = -log_{10}K_a = -log_{10}(1.7 \times 10^{-5}) = 4.77$ (3 s.f.)

2 Benzoic acid, a weak acid, has a pK_a value of 4.20. Calculate the value of the acid dissociation constant, K_a.

Answer

2 $pK_a = -log_{10}K_a$

$4.20 = -log_{10}K_a$

$K_a = 10^{-pK_a} = 10^{-4.20} = 6.31 \times 10^{-5}\ mol\ dm^{-3}$

Relating the pH of a weak acid to the dissociation constant K_a and the concentration

A weak acid is only partially dissociated and an equilibrium exists between the acid and its ions. The equilibrium for a weak acid is as follows:

$$HX(aq) \rightleftharpoons H^+(aq) + X^-(aq)$$

The expression for the acid dissociation constant, K_a, is given by

$$K_a = \frac{[H^+(aq)][X^-(aq)]}{[HX(aq)]}$$

Now $[H^+(aq)] = [X^-(aq)]$ because for each mole of H^+ there is a corresponding mole of X^-.

so $[H^+(aq)][X^-(aq)]$ can be replaced by $[H^+(aq)]^2$

Hence, $K_a = \frac{[H^+(aq)]^2}{[HX(aq)]}$

Rearranging this equation for $[H^+(aq)]^2$ gives

$$[H^+(aq)]^2 = K_a \times [HX(aq)]$$

If K_a and $[HX(aq)]$ are known, then $[H^+(aq)]^2$ can be found and by square rooting, so the value of $[H^+(aq)]$ is determined.

The value of $[H^+(aq)]$ is substituted into the equation $pH = -\log_{10}[H^+]$ to find the pH.

Example

Methanoic acid (HCOOH) is a weak acid and therefore dissociates only slightly in aqueous solution.

If the value of the acid dissociation constant K_a for methanoic acid at 25 °C is 1.78×10^{-4} mol dm^{-3}, calculate the pH of a 0.0455 mol dm^{-3} solution of this acid.

Answer

The equation for the equilibrium is:

$HCOOH(aq) \rightleftharpoons H^+(aq) + HCOO^-(aq)$

The expression for the acid dissociation constant, K_a is:

$$K_a = \frac{[H^+(aq)][HCOO^-(aq)]}{[HCOOH(aq)]}$$

Now, at equilibrium $[H^+(aq)] = [HCOO^-(aq)]$

Hence we can write:

$$K_a = \frac{[H^+(aq)]^2}{[HCOOH(aq)]}$$

$$1.78 \times 10^{-4} = \frac{[H^+(aq)]^2}{[0.0455]}$$

$$[H^+(aq)]^2 = 1.78 \times 10^{-4} \times 0.0455$$

$$= 8.099 \times 10^{-6}$$

$$[H^+(aq)] = \sqrt{8.099 \times 10^{-6}}$$

$$= 2.8459 \times 10^{-3} \text{ mol dm}^{-3}$$

$$pH = -\log_{10}[H^+] = -\log_{10}[2.8459 \times 10^{-3}] = 2.55 \text{ (3 s.f.)}$$

Pointer

At equilibrium, the number of moles of H^+ and $HCOO^-$ is the same. Hence, the concentrations of these ions will be the same as they are in the same volume of solution.

Pointer

The concentration, $[HCOO^-(aq)]$, has been swapped for $[H^+(aq)]$ in this expression.

Buffer action

Buffer solutions resist changes in the acidity or alkalinity of a solution when either H^+ or OH^- ions are added, thus keeping the pH almost constant.

An acidic buffer solution is made from a solution of a weak acid and a soluble salt of the acid that will keep the pH of the solution acidic (i.e. pH < 7).

Calculating the pH of acidic buffer solutions

An acidic buffer consists of a weak acid which is therefore only partially dissociated and exists in equilibrium with its ions:

$$HX(aq) \rightleftharpoons H^+(aq) + X^-(aq)$$

» Pointer

The concentrations of H^+ and X^- are equal and very small because HX is dissociated only partially.

When alkali is added, the OH^- ions react with the HX producing H_2O and X^-, according to the reaction:

$$HX(aq) + OH^-(aq) \longrightarrow H_2O(aq) + X^-(aq)$$

This has the effect of mopping up the OH^- ions so the pH of the solution remains almost constant.

When acid is added, the increased concentration of H^+ ions causes the equilibrium in the first reaction to shift to the left, thus increasing the concentration of HX. As the HX is not dissociated, this effectively removes the added H^+ ions and means that the pH of the buffer solution will remain constant.

Examples

1 Calculate the pH of the acidic buffer solution formed when 100 cm^3 of 0.250 mol dm^{-3} NaOH is added to 150 cm^3 of 0.400 mol dm^{-3} of a weak acid HX. The value of K_a for the weak acid HX is 4.55×10^{-5} mol dm^{-3}.

Answer

1 Number of moles of NaOH in 100 cm^3, $n = \frac{V}{1000} \times c$

$$= \frac{100}{1000} \times 0.250$$

$$= 0.025 \text{ mol}$$

» Pointer

NaOH is a strong alkali and so is completely dissociated so the concentrations of NaOH and OH^- will be equal.

Now, number of moles of NaOH = number of moles of OH^-

So number of moles of OH^- = 0.025 mol

Moles of HX initially present in 150 cm^3, $n = \frac{V}{1000} \times c$

$$= \frac{150}{1000} \times 0.40$$

$$= 0.060 \text{ mol}$$

Moles of HX present in buffer = moles of HX – moles of OH^-

$$= 0.060 - 0.025$$

$$= 0.035 \text{ mol}$$

$$[HX] = \frac{0.035}{250} \times 1000$$

$$= 0.14 \text{ mol dm}^{-3}$$

Number of moles of X^- in buffer = number of moles of OH^-

Number of moles of X^- in buffer = 0.025 mol

$$[X^-] = \frac{0.025}{250} \times 1000 = 0.1 \text{ mol dm}^{-3}$$

Expression for the equilibrium constant $K_a = \frac{[H^+][X^-]}{[HX]}$

Rearranging for $[H^+]$ gives $[H^+] = \dfrac{K_a \times [HX]}{[X^-]} = \dfrac{4.55 \times 10^{-5} \times 0.14}{0.1}$

$= 6.37 \times 10^{-5}$

$pH = -\log_{10}[H^+] = -\log_{10} 6.37 \times 10^{-5} = 4.20$ (3 s.f.)

2 Methanoic acid is a weak acid with a value of K_a at 25 °C of 1.78×10^{-4} mol dm^{-3}.

A buffer solution is prepared by mixing 1.84×10^{-2} mol of sodium methanoate with 2.35×10^{-2} mol of methanoic acid solution and the mixture is made up to 1.00 dm^3.

a) Calculate the pH of this buffer solution at 25 °C.

b) If 5.00 cm^3 of 0.100 mol dm^{-3} hydrochloric acid was added to the buffer solution described above, calculate the new pH of the solution.

Answer

» Pointer
As the weak acid is only very slightly ionised all the X^- ions can be considered to come from the ionisation of the sodium methanoate. Hence $[X^-]$ is equal to the concentration of the sodium methanoate.

2 a) $K_a = \dfrac{[H^+][X^-]}{[HX]}$

Rearranging for $[H^+]$ gives

$[H^+] = \dfrac{K_a[HX]}{[X^-]}$

$= \dfrac{1.78 \times 10^{-4} \times 2.35 \times 10^{-2}}{1.84 \times 10^{-2}}$

$= 2.27 \times 10^{-4}$ mol dm^{-3}

$pH = -\log_{10}[H^+] = -\log_{10}[2.27 \times 10^{-4}] = 3.64$ (3 s.f.)

Pointer
Note that the methanoate ions are the $HCOO^-$ ions.

b) The buffer solution combines the hydrogen ions added (i.e. from the HCl(aq)) with the methanoate ions from the sodium methanoate to form methanoic acid. This will result in the concentration of the methanoate ions falling and the concentration of the methanoic acid rising.

Number of moles of H^+ ions added $= \dfrac{cV}{1000}$

$= \dfrac{0.100 \times 5}{1000} = 5.00 \times 10^{-4}$ mol

» Pointer
Here we add the number of moles of the H^+ ions to the number of moles of methanoic acid. We also need to reduce the number of methanoate ions by this number.

Number of moles of methanoic acid

$= 2.35 \times 10^{-2} + 5.00 \times 10^{-4}$

$= 2.4 \times 10^{-2}$ mol

Number of moles of $HCOO^-$ ions

$= 1.84 \times 10^{-2} - 5.00 \times 10^{-4}$

$= 1.79 \times 10^{-2}$ mol

$[H^+] = \dfrac{K_a[HX]}{[X^-]}$

$= \dfrac{1.78 \times 10^{-4} \times 2.40 \times 10^{-2}}{1.79 \times 10^{-2}}$

$= 2.39 \times 10^{-4}$ mol dm^{-3}

$pH = -\log_{10}[H^+] = -\log_{10}[2.39 \times 10^{-4}] = 3.62$ (3 s.f.)

Test yourself 26

1. Calculate the pH of a 0.015 mol dm^{-3} solution of hydrochloric acid at 25 °C.
2. Calculate the pH of the solution formed when 25.0 cm^3 of 0.165 mol dm^{-3} hydrochloric acid are added to 975 cm^3 of water. Give your answer to three significant figures.
3. A solution has a pH of 5. Find the hydrogen ion concentration in the solution in mol dm^{-3}.
4. 4.00 g of hydrogen chloride was dissolved in water and the resulting solution was made up to 500 cm^3. Calculate the pH of the solution.
5. Calcium hydroxide is an example of a strong alkali. 4.0 g of calcium hydroxide, $Ca(OH)_2$, is dissolved in 500 cm^3 of water. Calculate the pH of the resulting solution to three significant figures.
6. Ethanoic acid (CH_3COOH) only dissociates slightly in aqueous solution.
   (a) (i) Write an equation for this dissociation.
   (ii) Write an expression for the dissociation constant K_a for ethanoic acid.
   (b) The value of K_a for ethanoic acid is 1.7×10^{-5} mol dm^{-3}. Calculate the pH of a 0.0254 mol dm^{-3} solution of ethanoic acid.

Chapter 27 Analytical techniques

27.1 Mass spectrometry

Mass spectroscopy can be used to find the relative molecular mass of an organic compound and give some information about the structure of the compound. A vaporised sample of the compound to be investigated is injected into a mass spectrometer where it is ionised and then separated out using electric and magnetic fields. The ions are separated out according to their mass to charge (m/z) ratio and a computer analyses the results.

When one electron is removed from the molecule, the molecular ion M^+ is produced and this has the highest m/z value and will be the right-most peak on the graph. Other ions are produced due to the breaking up of the compound into charged fragments with different m/z values.

Mass spectra

The results from mass spectroscopy are presented graphically with the m/z value along the x-axis and the relative abundance plotted on the y-axis. The heights of the peaks on the graph are shown as a percentage of the base peak with the base peak being the peak due to the molecular ion. For example, for methane, CH_4, the molecular ion will be CH_4^+ and this is the right-most peak with an m/z value of 16.

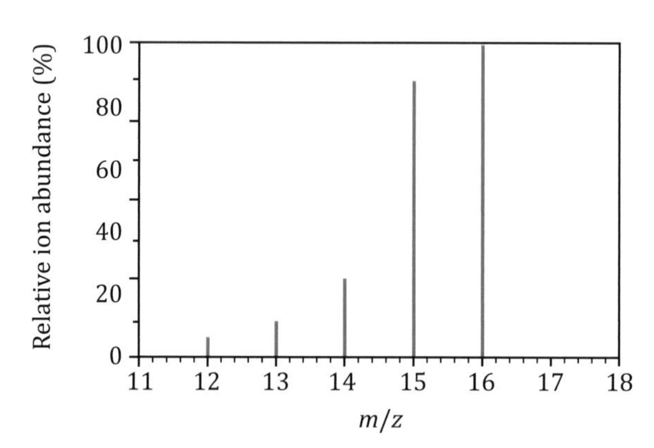

If you look at the above mass spectrum you can spot the molecular fragments CH_3^+ with an m/z value of 15, CH_2^+ with an m/z value of 14, CH^+ with an m/z value of 13 and C^+ with an m/z value of 12.

Example

The diagram below shows the mass spectrum for a sample of pure ethanol, CH_3CH_2OH:

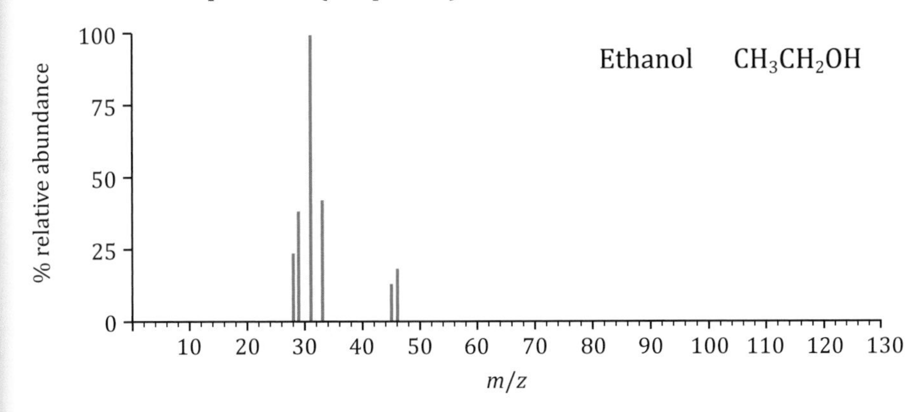

a) State the m/z value of the molecular ion M^+ and hence explain how this is used to determine the M_r of ethanol.

b) The commonest fragment ion has an m/z of 31. Identify the fragment ion causing this peak.

c) Give the m/z of the peak caused by the fragment ion $[CH_3CHOH]^+$.

Answer

a) M^+ (i.e. the fragment ion $[CH_3CH_2OH]^+$) is the right-most peak with $m/z = 46$. The M_r will be 46 as the m/z of the molecular ion is 46.

b) One way to approach this question is to find the mass of the fragment that has been removed by subtracting the m/z of the fragment ion from the m/z of the molecular ion. Here we have, $46 - 31 = 15$. We can now work out what might have been removed that has an m/z of 15. 15 is too small for an O to be removed as the m/z for O would be 16. Also 2 carbons could not have been removed as this would give 24 which is too much.

   Hence there must be a combination of C and H atoms. CH_3 has an m/z value of 15.

   Hence the peak represents the loss of the methyl group (i.e. CH_3) from the molecule to form the fragment ion $[CH_2OH]^+$.

   $[CH_2OH]^+$ has an m/z of $12 + 2 + 16 + 1 = 31$.

c) The fragment ion $[CH_3CHOH]^+$ will have an m/z value of $12 + 3 + 12 + 1 + 16 + 1 = 45$.

27.2 Infrared spectroscopy

When radiation having a range of infrared frequencies is shone through a sample of an organic compound it is found that some of the frequencies are absorbed by the sample. This results in some frequencies being missing from the radiation when it emerges from the sample. The frequencies that are absorbed are used to stretch or bend bonds in the molecules. It is possible to use these missing frequencies to give information about the different types of bonds in the sample.

An infrared spectrometer produces a graph called an infrared spectrum with the transmittance (i.e. the percentage of the infrared radiation getting through the sample) plotted against the frequency.

Instead of using the usual units for frequency of Hertz (Hz), in spectroscopy we use another unit, called the wavenumber, to represent the frequency, which has the units cm^{-1}.

Table of infrared absorption data

In questions you will be asked to interpret infrared spectra and you will need to use a table of absorption data similar to the following:

Bond	Wavenumber/cm^{-1}
C—C	750–1100
C—O	1000–1300
C=C	1620–1680
C=O	1680–1750
C≡N	2220–2260
O—H (in acids)	2500–3000
C—H	2850–3300
O—H (in alcohols)	3230–3550
N—H (in amines)	3300–3500

Take a look at the following infrared spectrum for ethanoic acid:

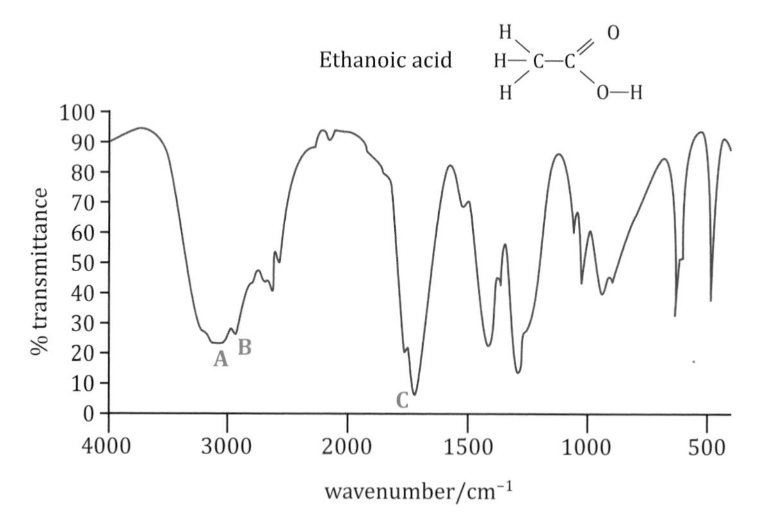

If you look at the structural formula of ethanoic acid shown on the graph you can see there are the following types of bond.

C—C

C—H

C=O

O—H

C—O

Now we can use the table of data to which of the bonds are represented by A, B and C marked on the graph.

Pointer

Always draw the structural formula of the molecule so you can see all the bonds that are present, some of which can be identified on the infrared spectrum.

The broad peak at A, with a wavenumber of around 3000 cm^{-1}, is characteristic of an O—H group in an acid (according to the table this peak is at 2500–3000 cm^{-1}).

The smaller peak at B, at around 2900 cm^{-1} which is masked by the wider O—H group, is characteristic of the C—H bonds (according to the table this peak is at 2850–3300 cm^{-1}).

The peak at C at around 1700 cm^{-1} is characteristic of the C=O bond (according to the table this peak is at 1680–1750 cm^{-1}).

Example

The infrared spectrum for ethanol is shown below:

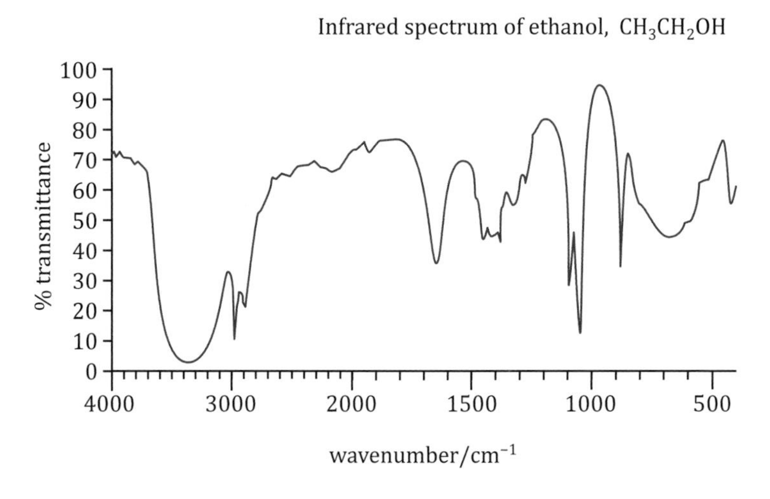

Using the data from the table on page 166, give the wavenumber that corresponds to each of the following bonds:

a) O—H bond in an alcohol

b) C—H bond

c) C—O bond

Answer

a) 3350 cm^{-1}

b) 2950 cm^{-1}

c) 1050–1100 cm^{-1}

Pointer

You need to refer to the table of absorption data which is included on page 166. Make sure you look carefully at the data booklet used by your particular specification as the way in which the data is presented differs slightly.

Using infrared spectra to identify structural isomers

The two infrared spectra shown on page 168 are for two structural isomers having the molecular formula C_6H_{12}.

If we are told that one of the compounds is unsaturated and the other is saturated, then we will be able to distinguish to which compound each spectra belongs.

Spectrum A

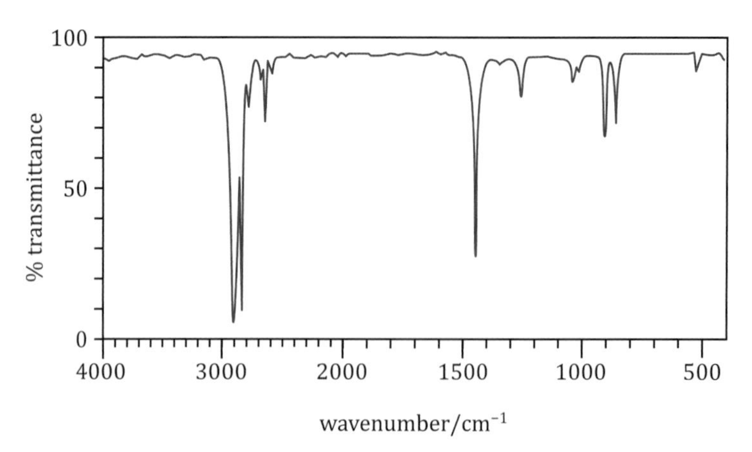

Spectrum B

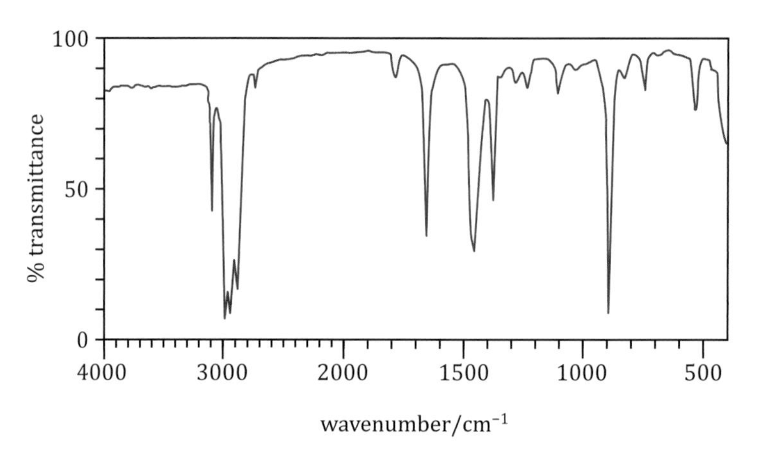

The unsaturated compound (i.e. containing a C=C bond), will have a peak in the 1620–1680 cm^{-1} range according to the table. Looking at the two spectra you can see that there is no peak in this region for spectrum A but there is in spectrum B. Hence the compound represented by spectrum B is the unsaturated compound and that represented by spectrum A is the saturated compound.

Example

An organic compound was analysed and thought to have the following structure:

```
     OH H  OH H  OH
     |  |  |  |  |
  H—C—C—C—C—C—C=O
     |  |  |  |  |  |
     H  OH H  OH H  H
```

Infrared spectroscopy was used to check if this was the correct structure.

Using the infrared spectrum for the compound, explain with reasons why the above cannot be the correct structure.

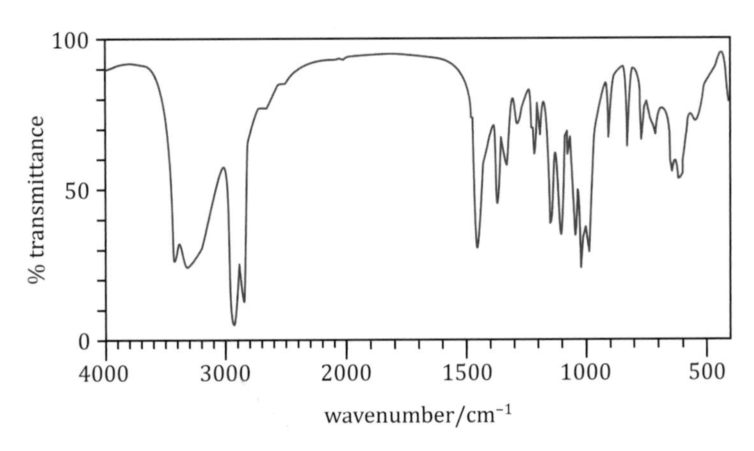

Note you will need to refer to the table on page 166 to complete this question.

Answer

You need to first identify the functional groups/bonds present in the compound.

There are the following bonds present:

C—C

C—H

O—H

C=O

Almost all organic compounds have C—H and C—C bonds, so these bonds do not give much information.

The broad peak around 3300 cm^{-1} is produced by an O—H bond in an alcohol.

There should be a peak at 1680–1750 cm^{-1} due to the C=O but there is no peak here in the infrared spectrum. Hence the structure with the C=O bond must be incorrect.

Pointer

You need to refer to the table of absorption data which is included on page 166. Make sure you look carefully at the data booklet used by your particular specification as the way in which the data is presented differs slightly.

27.3 Nuclear magnetic resonance (NMR) spectroscopy

In nuclear magnetic resonance spectroscopy, certain nuclei absorb low frequency electromagnetic radiation when a magnetic field is applied and this causes such nuclei to move to a higher energy state and resonate.

Organic compounds contain hydrogen atoms each containing a proton. These protons are shielded from the applied magnetic field by the surrounding electrons. As many hydrogen atoms are in different environments depending on the groups they are next to, the amount of shielding depends on the bond polarity and if there are electron-donating or electron-withdrawing groups in the compound.

Pointer

Note that sometimes you will see protons referred to and sometimes hydrogens. In NMR spectroscopy we use these interchangeably.

There are two types of NMR spectra. Low resolution spectra where a series of peaks is seen without any splitting, and high resolution spectra where the peaks have been split into a series of smaller peaks grouped around where the single peak would be situated.

As many of the hydrogen atoms in an organic compound are in different environments, the type of resonance can be used to identify the structure of a compound.

Take methane, for example, each hydrogen atom is in the same environment so they will resonate at the same frequency. However, in ethanol (CH_3CH_2OH), the protons in CH_3 and CH_2 and OH are in different environments and will resonate at different frequencies.

Changes in shielding cause changes in frequency which is measured by a quantity called chemical shift and this has the units ppm (parts per million).

Chemical shift is a relative quantity and is measured relative to an organic compound called tetramethylsilane, called TMS for short. The chemical shift value δ is defined as being zero and nearly all proton absorptions are on a scale of 0–10 below this value.

Look at the following NMR spectrum for the compound diethyl ketone:

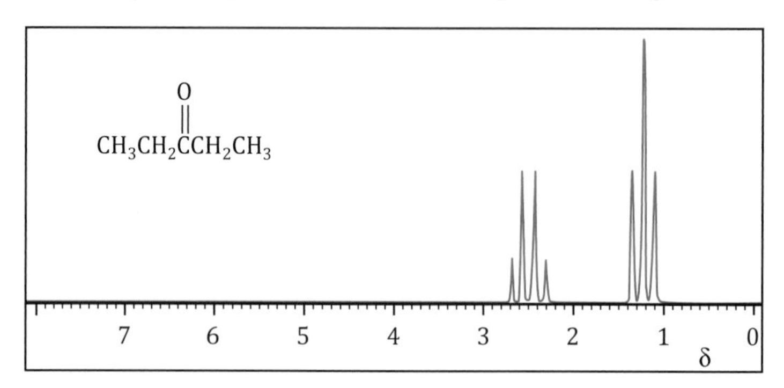

You can see the following from the spectrum:

There are two absorptions (as there are two groups of peaks) – this shows that there are two non-equivalent protons present (i.e. one for the protons in CH_2 and the other for the protons in CH_3).

The splitting of each absorption can be used to give information about neighbouring protons. In this spectrum, the three peaks, called a triplet, are caused by the peak due to the equivalent hydrogens in the methyl group being split by the two hydrogens on the adjacent carbon. In a similar way the quartet is caused the by the hydrogens in the CH_2 group being split by the neighbouring CH_3 group. Notice that because of the symmetry of the compound, the hydrogens in the methyl groups at the end of the molecule are in identical environments as are the two CH_2 groups.

Predicting the splitting

If there are n hydrogens on an adjacent carbon then this will split the peak into $n + 1$ smaller peaks. For example, the compound ethanal, CH_3CHO has the structural formula:

```
     H
     |    O
     |  //
  H—C—C
     |   \
     |    H
     H
```

There are two different environments for the hydrogens. The hydrogens on the methyl group are all in the same environment, whilst the hydrogen forming the CHO group is in a different environment. This means that there would be two main peaks.

Spin-spin coupling is caused by the magnetic field on a hydrogen being affected by hydrogens on adjacent carbon atoms. Spin-spin coupling is responsible for peaks being split into several peaks near to each other.

In ethanal, spin-spin coupling causes the splitting of the main peaks into smaller peaks. Looking at the carbon of the methyl group there is one adjacent carbon to which a single hydrogen is attached. The general rule is that if there are *n* adjacent hydrogens, then the peak due to the hydrogen being considered will be split into $n + 1$ smaller peaks. This is called the $n + 1$ rule and it is used to predict the peak splitting in molecules.

Applying the $n + 1$ rule to the single hydrogen in ethanal. This will split the peak due to the methyl hydrogens into two smaller peaks. In a similar way, the 3 hydrogens on the methyl group will split the peak due to the single hydrogen on the adjacent carbon into 3 + 1 = 4 peaks.

Using the data obtained from the table of ^{1}H NMR chemical shift data we can determine where the peaks will be situated.

According to the data in the table on page 172, the hydrogen in the aldehyde group R-C**H**O (note the hydrogen we are referring to is made bold) has a value of 9.0–10.0 ppm. Hence there will be a quartet (i.e. 4 small peaks) somewhere in this range.

The hydrogens in the methyl group will form a doublet with a value of 2.1–2.6. Here we look for a hydrogen on a carbon adjacent to a carbon containing a carbonyl group (i.e. C=O).

The heights of the peaks

Splitting of a peak into smaller peaks can result in these smaller peaks being of different heights.

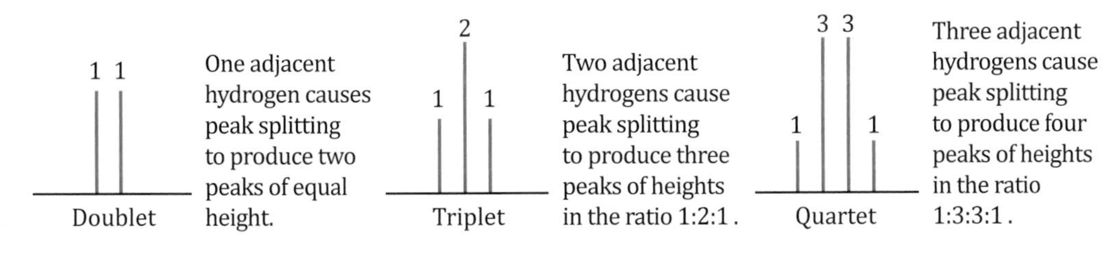

Pointer

This singlet is produced by all the equivalent hydrogens on a methyl group attached to the carbonyl group.

Table of ^{1}H NMR chemical shift data

Type of proton	δ/ppm
ROH	0.5–5.0
RCH_3	0.7–1.2
RNH_2	1.0–4.5
R_2CH_2	1.2–1.4
R_3CH	1.4–1.6
R—C(=O)—C(—)(—**H**)—	2.1–2.6
R—O—C(—)(—**H**)—	3.1–3.9
RCH_2Cl or Br	3.1–4.2
R—C(=O)—O—C(—)(—**H**)—	3.7–4.1
R(—)C=C(—)**H**	4.5–6.0
R—C(=O)—**H**	9.0–10.0
R—C(=O)—O—**H**	10.0–12.0

Notice the way the hydrogen that causes the peak is emboldened in this table.

The areas of the peaks

Take ethanol (CH_3CH_2OH), for example. There are three times the number of hydrogens in the methyl group compared with hydrogens in the OH group. This results in the area of the peak for the methyl group being three times the area of the peak for the OH group.

How to predict the number of peaks split by adjacent hydrogens

One hydrogen on the adjacent carbon – the peak is split into two with each having the same height.

Two hydrogens on the adjacent carbon – the peak is split into three (called a triplet) with height ratio 1:2:1

Three hydrogens on the adjacent carbon – the peak is split into four (called a quartet) with height ratio 1:3:3:1

Example

The ketone, butan-2-one has the following structural formula:

```
    H  H  O  H
    |  |  ‖  |
 H—C—C—C—C—H
    |  |     |
    H  H     H
```

a) How many different environments are there for hydrogens in this compound?

b) There is a methyl singlet at around 2.2 ppm in the NMR spectra for this compound.

   i) Copy the structural formula and draw a ring around any one of the hydrogens that are responsible for this singlet.

   ii) Explain why a singlet is produced.

c) The hydrogens in the CH_2 group are labelled H_a in the structure shown below.

```
    H  Ha O  H
    |  |  ‖  |
 H—C—C—C—C—H
    |  |     |
    H  Ha    H
```

   i) What range of values would the hydrogens in the CH_2 have in the NMR spectrum?

   ii) Explain why the peak due to these hydrogens is split into a quartet of smaller peaks.

Answer

a) There are three different environments.

To see this you are best drawing a diagram showing the bonds:

```
    H  H  O  H
    |  |  ‖  |
 H—C—C—C—C—H
    |  |     |
    H  H     H
```

Three different environments for hydrogens

b) i) Looking up 2.2 ppm in the table gives the following:

R—C(=O)—C(H)— (O on the C=O carbon; **H** on the adjacent carbon)	2.1–2.6

A circle can be drawn around any of the hydrogens in the methyl group attached to the carbonyl group, like this:

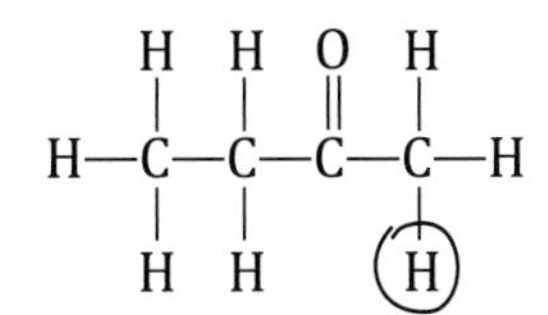

» *Pointer*

This singlet is produced by all the equivalent hydrogens on a methyl group attached to the carbonyl group.

ii) A singlet is produced because there are no hydrogens on the adjacent carbon (i.e. the carbon forming the carbonyl group). This means there is no spin-spin coupling and therefore no splitting of the peak.

c) i) 2.1–2.6 ppm (this value is obtained from the table).

ii) Spin-spin coupling due to the three hydrogens in the methyl group will cause splitting of the peak into a quartet of smaller peaks.

NMR spectra using carbon-13

Carbon-12 has no nuclear spin and therefore cannot be detected by NMR. Carbon-13 does have nuclear spin and even though only 1 in 100 carbon atoms are C-13 (i.e. ^{13}C), modern NMR spectrometers are able to produce a C-13 NMR spectrum.

Like hydrogen atoms, carbon atoms in different functional groups are shielded differently and produce a chemical shift ppm; however, because of the very low concentrations of C-13 no spin-spin coupling occurs so the peaks are not split into smaller peaks.

The following table shows the chemical shift values for C-13 in different functional groups:

Table of ^{13}C NMR chemical shift data

Type of carbon	δ/ppm
—C—C—	5–40
R—C—Cl or Br	10–70
R—C(=O)—C—	20–50
R—C—N<	25–60
—C—O—	50–90
>C=C<	90–150
R—C≡N	110–125
(benzene ring)	110–150
R—C(=O)—	160–185
R—C(=O)—	190–220

Examples

1 Compounds A and B have the molecular formula $C_4H_{11}N$. If both compounds are secondary amines and in the ^{13}C NMR spectra, compound A gave two peaks whilst compound B gave three peaks, draw possible structures for each compound.

Answer

1 As both compounds are secondary amines they must both have a structure of the form:

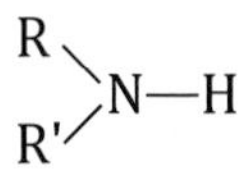

R(R')N—H

where R and R' could be the same or different groups.

Compound A gave two peaks so there must be two different environments for the carbons. By deduction, the alkyl groups R and R' must be the same giving the following structure:

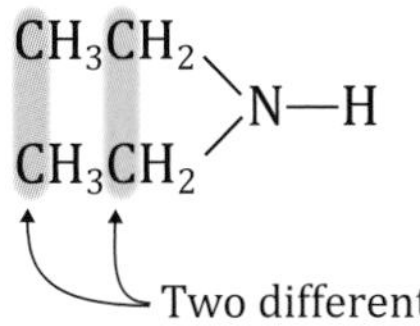

$(CH_3CH_2)(CH_3CH_2)N—H$

Two different environments for carbon, so there are 2 peaks

Compound B gave three peaks so there are three different environments for the carbon:

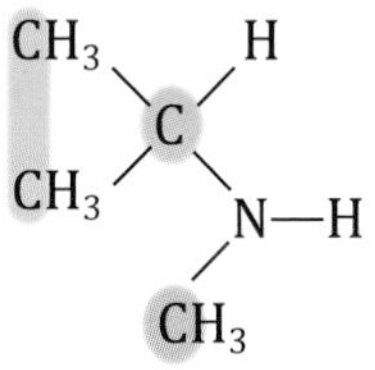

$(CH_3)(CH_3)C(H)—N(H)—CH_3$

Three different environments for carbon, so there are 3 peaks.

2 Three samples of organic compounds have been become mixed up. The three samples were:

1,4-dibromobutane

Butane-1,4-diol

Butanedioic acid

A sample of each compound was examined using infrared spectroscopy and the following spectra were obtained.

Examine each spectrum using the table of data on page 166 and state with reasons which compound from the above list each sample represents.

» Pointer

Don't be afraid of drawing incorrect structures first but remember to cross them out when you find the correct one so the examiner knows which one to mark.

Sample A

Butanedioic acid

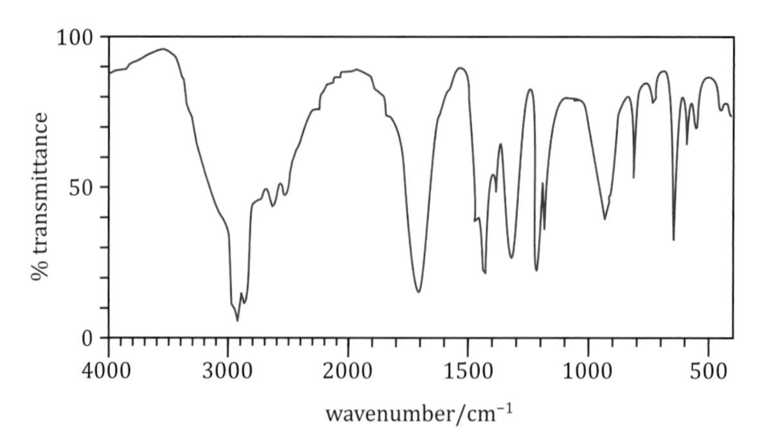

Sample B

Butane-1,4-diol

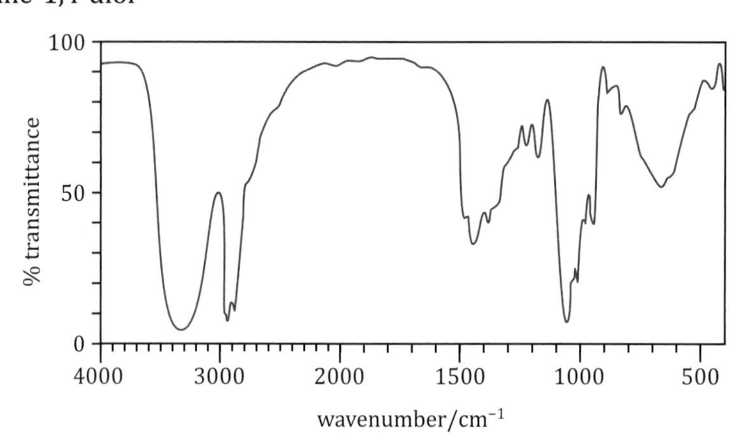

Sample C

1,4-dibromobutane

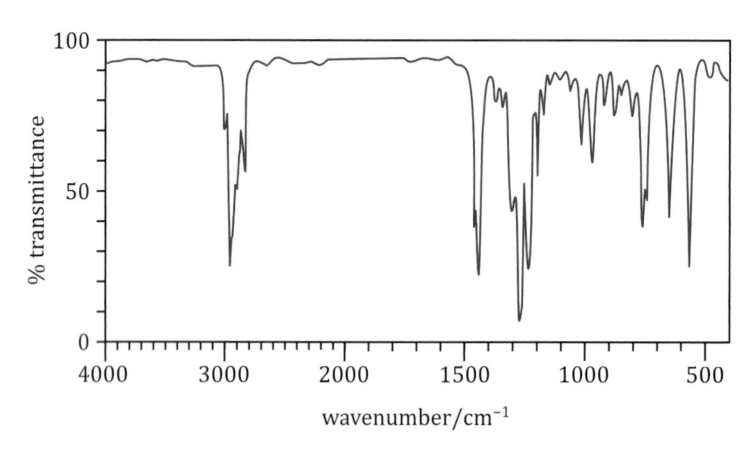

Answer

2 The first thing to do is to write down the structural formulae of all three compounds in the list:

```
   Br H  H  Br
   |  |  |  |
H—C—C—C—C—H      1,4-dibromobutane
   |  |  |  |
   H  H  H  H
```

```
   OH H  H  OH
   |  |  |  |
H—C—C—C—C—H      Butane-1,4-diol
   |  |  |  |
   H  H  H  H
```

```
   OH H  H  OH
   |  |  |  |
O=C—C—C—C=O      Butanedioic acid
      |  |
      H  H
```

In butanedioic acid there are two carbonyl groups (i.e. C=O), which according to the table of data would have peaks at 1680–1750 cm^{-1}. Sample A has a peak at 1700 cm^{-1} and as none of the other compounds have a carbonyl group, sample A must be butanedioic acid.

Butane-1,4-diol contains two OH groups in a alcohol. According to the table, OH groups in an alcohol have a peak at 3230–3550 cm^{-1}. Sample B has a wide peak at 3300 cm^{-1} so it must be the alcohol butane-1,4-diol.

By elimination, sample C must be 1,4-dibromobutane.

3 Chloropropane C_3H_7Cl exists as two isomers. Sketch the low resolution NMR spectrum of each of the isomers and mark on your diagram the approximate chemical shift in ppm and the relative area of each peak.

Answer

3 We first draw the structural formulae of the two isomers 1-chloropropane and 2-chloropropane.

```
   H  H  Cl
   |  |  |
H—C—C—C—H      1-chloropropane
   |  |  |
   H  H  H
```

```
   H  Cl H
   |  |  |
H—C—C—C—H      2-chloropropane
   |  |  |
   H  H  H
```

For $CH_3CH_2CH_2Cl$ there are three different environments for hydrogens resulting in three peaks. There will be a peak due to the hydrogens in a $RC\mathbf{H}_3$ at 0.7 to 1.2 ppm, a peak due to hydrogens in a $RC\mathbf{H}_2Cl$ at 3.1 to 4.2 ppm and a peak due to hydrogens in an $RC\mathbf{H}_2$ at 1.2 to 1.4 ppm. The areas of these peaks will be in the ratio 3 (for the methyl group) to 2 to 2 as the other two carbons both contain two hydrogens each.

» Pointer

As this question is concerned with low-resolution NMR spectra you do not need to include any peak splitting as this only shows up on high-resolution spectra.

Adding this data to a sketch gives the following:

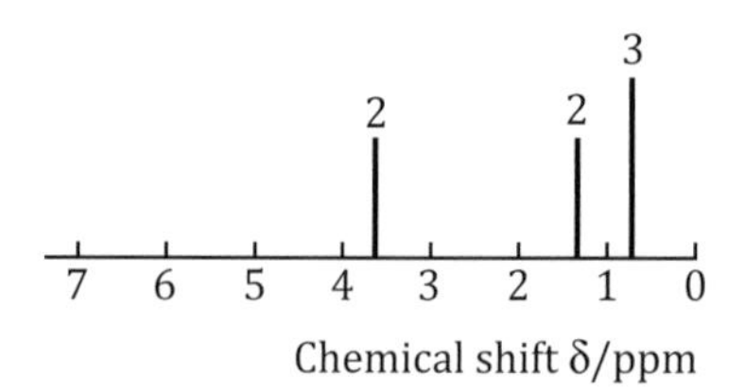

For $(CH_3)_2CHCl$, there are two different environments for the hydrogens. The hydrogens in the two methyl groups at either end of the molecule will form one peak while the hydrogen attached to the carbon in the middle of the molecule will form the other peak.

Using the data in the table we can identify that there will be a peak due to a hydrogen in a RC**H**$_2$Cl at 3.1 to 4.2 ppm and a peak due to hydrogens in a R_2CH_2 at 1.2 to 1.4 ppm.

As there are 6 equivalent hydrogens in the two methyl group and only one hydrogen attached to the carbon in the middle, the areas of the peaks will be in the ratio of 6 : 1.

With this information, the following sketch can be produced:

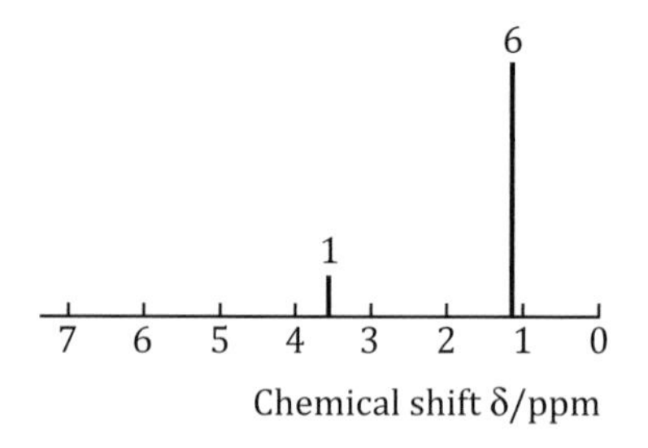

27.4 The relationship between energy and frequency

When an electron becomes excited it receives energy and if the energy is sufficient, it can be promoted to higher energy levels. In some cases the promotion of electrons to higher energy levels can be achieved using visible light. Most transition metal compounds are coloured because the light absorbed by them is in the visible part of the electromagnetic spectrum. It is the combination of the colours remaining from the white light that you see. If you think, in simple terms, that white light is made up from the colours red, blue and green and the red light is absorbed by a certain compound then the blue and green are transmitted to give a greenish blue colour to the compound.

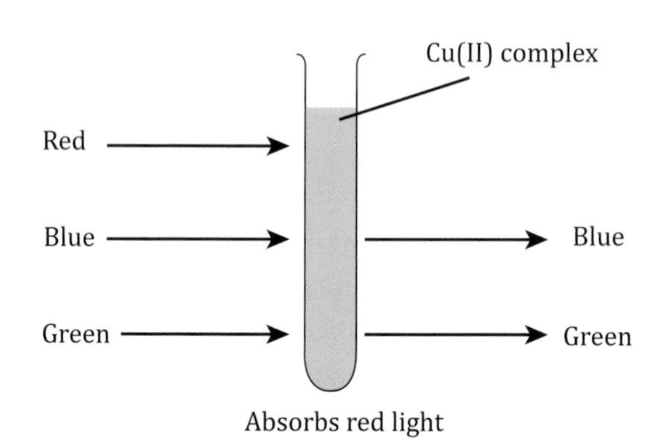

The relationship between frequency, wavelength and speed of light

This relationship is given by $c = f\lambda$ where c is the speed of light (i.e. a constant value of 3×10^8 m s^{-1}), f is the frequency in Hz and λ is the wavelength in m. As f increases, λ decreases in order to keep the speed of light c at its constant value. If f decreases, λ increases in order to keep the speed of light c at its constant value.

Transition metal ions (such as Cu^{2+} which is blue) can be identified by their colour. Colour arises when some of the wavelengths of visible light are absorbed and the remaining wavelengths of light are transmitted or reflected. d electrons move from the ground state to an excited state when light is absorbed. The colour you see in a transition metal compound is the visible light that is not absorbed by the 3d electronic transitions.

The following diagram shows the wavelengths that are absorbed by the aqueous copper(II) complex. Note that the light in the red-orange–yellow part of the spectrum is absorbed but the light in the green and blue parts of the spectrum is transmitted so the complex appears blue–green.

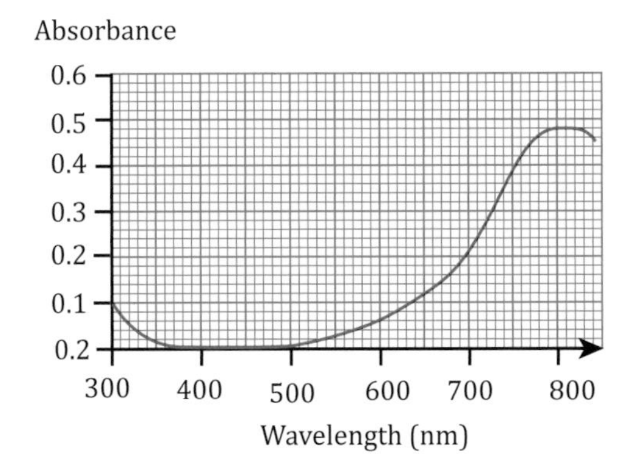

The energy difference between the ground state and the excited state of the d electrons is given by:

$$\Delta E = hf = \frac{hc}{\lambda}$$

(note that the equation $c = f\lambda$ is rearranged as $f = \frac{c}{\lambda}$)

where ΔE is the difference in energy levels in J,

f is the frequency of the light absorbed in Hz,

c is the speed of light in ms^{-1} (note that this is a constant value),

λ is the wavelength of the light in m,

h is Plank's constant (i.e. 6.63×10^{-34} J s).

(Note that in some syllabuses you will see the letter ν being used for frequency in this formula rather than f.)

Important note

The wavelength of light is often expressed in nanometres (nm) where 1 nm = 1×10^{-9} m.

Example

The following table shows the wavelengths of light and the colours corresponding to these wavelengths.

Wavelength (nm)	Colour of light
400	violet
450	blue
500	green
600	yellow
650	red

a) Red light is absorbed by a transition metal complex. Using the following data and the data in the table above, determine the amount of energy when **one** photon of **red** light is absorbed by an electron.

Planck's constant, $h = 6.63 \times 10^{-34}$ J s

Speed of light, $c = 3.00 \times 10^{8}$ m s^{-1}

b) A different transition metal complex absorbs light of different wavelengths, as shown in the following diagram.

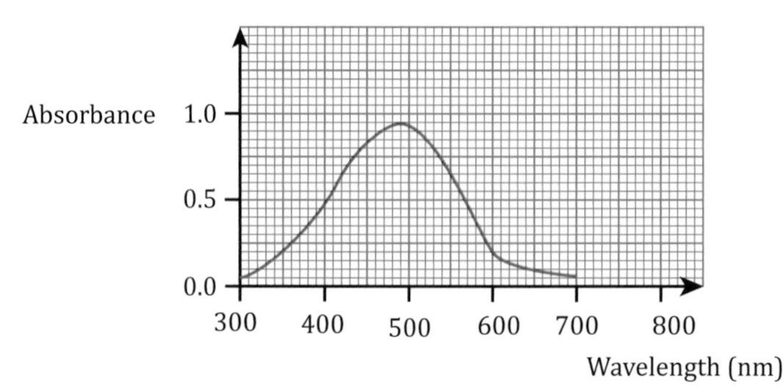

Here is a list of possible colours for this complex.

Blue Green Red Yellow

Choose **one** of the colours that best describes the colour of this complex.

Answer

a) $E = hf = hc/\lambda = \dfrac{6.63 \times 10^{-34} \times 3.00 \times 10^{8}}{650 \times 10^{-9}} = 3.06 \times 10^{-19}$ J

b) Inspection of the graph reveals that most light is absorbed in the region 400–600 nm. Hence, using the table for the wavelengths, the colours violet, blue, green and yellow are absorbed. Almost no light is absorbed at 650 nm so the colour of the complex will be red.

Test yourself 27

1. By referring to the table on page 166, state what structural feature in propanone and propanal can be seen in the infrared spectra by the absorptions at wavenumber 1720 cm^{-1}.

2. Student A reckons that the infrared spectrum shown here is that of an alkene, whereas student B reckons it is of an alkane. Use the data in the table on page 166 to say with a reason which student is correct.

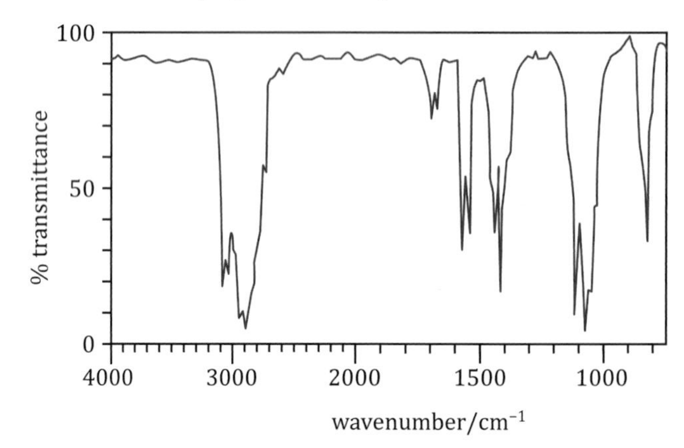

3. There are three isomers of $C_6H_4(NO_2)_2$ and these are shown here:

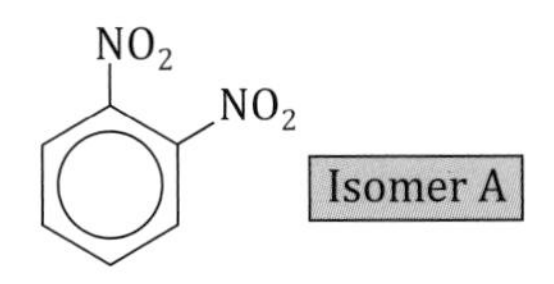

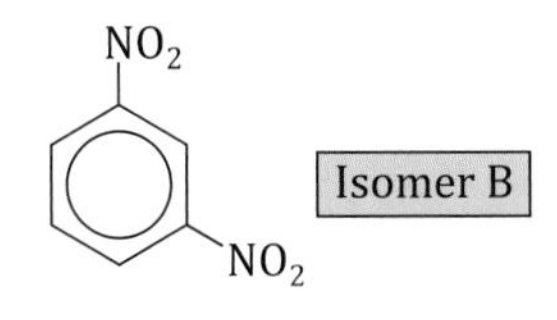

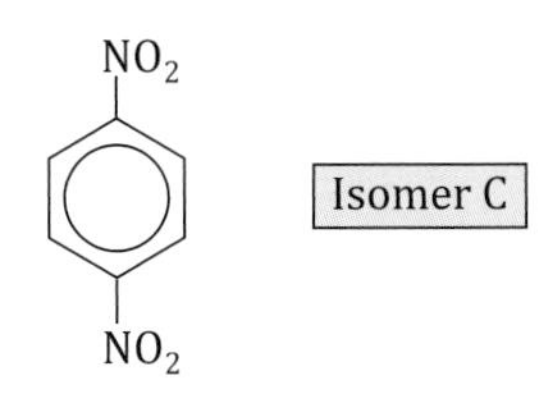

State with reasons the number of peaks in the ^{13}C NMR spectrum of each isomer.

4. Two compounds X and Y have the molecular formula C_5H_{12}. In their NMR spectra, X has only one peak whereas Y has three. Using this information, draw possible structural formulae for X and Y.

5. Compounds A and B have the molecular formula C_3H_6O. When examined using infrared spectroscopy, compound A had a peak at wavenumber 1720 cm^{-1} and compound B had two absorptions at wavenumbers 1645 cm^{-1} and 3300 cm^{-1}. Compound A had only one peak in its ^1H NMR spectrum.

   Using this information deduce one possible structure for each of the compounds A and B.

6. The ^1H NMR spectrum of 1,1-dibromoethane is shown below:

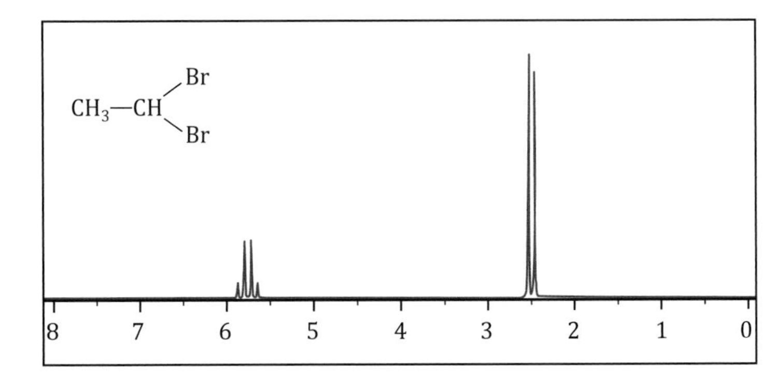

Assume that the bromine atoms do not contribute to any splitting.

By referring to the spectra and the structural formula of 1,1-dibromoethane explain the origin of each of the following.

(a) The doublet at 2.5 ppm.

(b) The quartet at 5.8 ppm.

Test yourself answers

1 Arithmetic and numerical computation

1.1
a) $3 + 14 \div 2 = 3 + 7 = 10$
b) $6 \times 3 - 4 \div 2 = 18 - 2 = 16$
c) $(5 \times 12) - 15 = 60 - 15 = 45$
d) $\frac{15 + 5}{2} = \frac{20}{2} = 10$
e) $18 - 4 \times 3 = 18 - 12 = 6$
f) $23.6 \times 3.1 - 9.8 = 73.16 - 9.8 = 63.36$
g) $(3 + 4)^2 = 7^2 = 49$
h) $4 + 5 \times 10^2 = 4 + 5 \times 100 = 504$
i) $\frac{5 \times 12}{30 \div 5} = \frac{60}{6} = 10$
j) $7 - \frac{18}{3} = 7 - 6 = 1$
k) $\frac{5}{10 + 15} = \frac{5}{25} = \frac{1}{5}$

1.2
a) 15.72
b) 39.55
c) 58.654

1.3
a) −1
b) −6
c) −1
d) 3
e) −5
f) 1
g) −3
h) −2
i) −1
j) −11

1.4
a) −4
b) 0
c) 2
d) 1
e) −5
f) −2
g) −4
h) 2

1.5
a) 6
b) −4
c) −3
d) −12
e) −3
f) −5
g) 3
h) −2
i) −4
j) 9
k) 2
l) 18

1.6
a) $\frac{\frac{3}{10}}{\frac{9}{25}} = \frac{3}{10} \div \frac{9}{25} = \frac{3}{10} \times \frac{25}{9} = \frac{1}{2} \times \frac{5}{3} = \frac{5}{6}$
b) $16 \div \frac{4}{5} = 16 \times \frac{5}{4} = 4 \times 5 = 20$
c) $\frac{24}{\frac{8}{15}} = 24 \times \frac{15}{8} = 3 \times 15 = 45$
d) $\frac{\frac{4}{5}}{8} = \frac{\frac{4}{5}}{\frac{8}{1}} = \frac{4}{5} \times \frac{1}{8} = \frac{1}{5} \times \frac{1}{2} = \frac{1}{10}$
e) $\frac{\frac{3}{25}}{3^2} = \frac{\frac{3}{25}}{9} = \frac{\frac{3}{25}}{\frac{9}{1}} = \frac{3}{25} \times \frac{1}{9} = \frac{1}{25} \times \frac{1}{3} = \frac{1}{75}$

1.7
a) Yes
b) Yes
c) No
d) No
e) Yes
f) Yes
g) No
h) Yes
i) Yes
j) No

1.8
a) $10.5 = 1.05 \times 10$
b) $200 = 2 \times 10^2$
c) $308.6 = 3.086 \times 10^2$
d) $1230 = 1.23 \times 10^3$
e) One million = 1 000 000 = 1×10^6
f) $0.3 = 3 \times 10^{-1}$
g) $0.00045 = 4.5 \times 10^{-4}$
h) $0.00000003 = 3 \times 10^{-8}$

i) $0.005 = 5 \times 10^{-3}$
j) $50.5 = 5.05 \times 10$

1.9 a) $15 \times 10^{-4} = 1.5 \times 10 \times 10^{-4} = 1.5 \times 10^{-3}$
b) $18 \times 10^{4} = 1.8 \times 10 \times 10^{4} = 1.8 \times 10^{5}$
c) $10 \times 10^{3} = 1 \times 10^{4}$
d) $125 \times 10^{-3} = 1.25 \times 10^{2} \times 10^{-3} = 1.25 \times 10^{-1}$
e) $0.1 \times 10^{-3} = 1 \times 10^{-1} \times 10^{-3} = 1 \times 10^{-4}$
f) $0.001 \times 10^{3} = 1 \times 10^{-3} \times 10^{3} = 1$
g) $0.5 \times 10^{4} = 5 \times 10^{-1} \times 10^{4} = 5 \times 10^{3}$
h) $0.125 \times 10^{-3} = 1.25 \times 10^{-1} \times 10^{-3} = 1.25 \times 10^{-4}$
i) $18.5 \times 10^{-2} = 1.85 \times 10 \times 10^{-2} = 1.85 \times 10^{-1}$
j) $0.002 \times 10^{4} = 2 \times 10^{-3} \times 10^{4} = 2 \times 10$

1.10 a) 6×10^{14}
b) 9×10^{5}
c) 1.6×10^{4}
d) 1.1×10^{-5}
e) 2×10^{2}
f) 2×10^{-7}
g) 2×10^{-2}
h) 1×10^{-2}
i) 9×10^{-3}
j) 1×10^{5}

1.11 a) 1:3
b) 1:3
c) 3:2
d) 3:1
e) 1:3
f) 5:12
g) 1:100
h) 1:4
i) 1:5
j) 1:5

1.12 a) 250 g : 2 kg = 250 g : 2000 g = 1:8
b) 1 mg : 10 g = 0.001 g : 10 g = 1:10000
c) 20 mg : 1 g = 0.02 g : 1 g = 0.02:1
d) 25 g : 250 g = 1:10
e) 400 mg : 20 g = 0.4g : 20 g = 1:50
f) 20 g : 1 kg = 20 g : 1000 g = 1:50
g) 4 mole : 8 mole = 1:2
h) 0.01 mole : 1 mole = 1:100
i) 0.2 mole : 0.01 mole = 20:1
j) 0.8 mole : 0.2 mole = 4:1

1.13 a) i) $Mg(s) + 2HCl(aq) \longrightarrow MgCl_2(aq) + H_2(g)$
1 mol 2 mol 1 mol 1 mol
Multiplying by 4 to find the amounts for 4 mol of $MgCl_2$ gives
4 mol 8 mol 4 mol 4 mol
Hence 4 mol of Mg are needed.

ii) $Mg(s) + 2HCl(aq) \longrightarrow MgCl_2(aq) + H_2(g)$
1 mol 2 mol 1 mol 1 mol
0.2 mol 0.4 mol 0.2 mol 0.2 mol
Hence 0.4 mol of HCl are needed.
Multiply by 0.2 to find the amounts for 0.2 moles of H_2.

b) i) $4Al(s) + 3O_2(g) \longrightarrow 2Al_2O_3(s)$
4 mol 2 mol
1 mol 0.5 mol
Divide by 4 to find the amount for 1 mole of Al.

ii) $4Al(s) + 3O_2(g) \longrightarrow 2Al_2O_3(s)$
3 mol 2 mol
0.75 mol 0.5 mol
We can save time by not writing the moles for the O_2 as we are not asked about them in the question. Divide the amounts by 4 to find the amounts for 0.5 mol of Al_2O_3.

2 Handling data

2.1 a) 10.6
b) 123.977
c) 0.03
d) 3.971
e) 0.105
f) 0.002
g) 4.10

2.2 a) 12.6666 = 12.7 (3 s.f.)
b) 0.06356 = 0.0636 (3 s.f.)
c) 1.87888 = 1.88 (3 s.f.)
d) 0.0008959 = 0.000896 (3 s.f.)
e) $1.2058 \times 10^{8} = 1.21 \times 10^{8}$ (3 s.f.)
f) $2.03988 \times 10^{-4} = 2.04 \times 10^{-4}$ (3 s.f.)

2.3
1. A
2. C
3. B
4. B
5. A
6. B
7. C
8. A
9. B
10. A

2.4

1. a) $5\ cm^3 = \frac{5}{1\,000\,000}\ m^3 = 5 \times 10^{-6}\ m^3$

   b) $250\ cm^3 = \frac{250}{1\,000\,000}\ m^3 = 2.5 \times 10^{-4}\ m^3$

   c) $0.5\ cm^3 = \frac{0.5}{1\,000\,000}\ m^3 = 5 \times 10^{-7}\ m^3$

   d) $1500\ cm^3 = \frac{1500}{1\,000\,000}\ m^3 = 1.5 \times 10^{-3}\ m^3$

2. a) $20\ dm^3 = 20 \times 1 \times 10^{-3}\ m^3 = 2 \times 10^{-2}\ m^3$
   b) $450\ dm^3 = 450 \times 1 \times 10^{-3}\ m^3 = 4.5 \times 10^{-1}\ m^3$
   c) $1800\ dm^3 = 1800 \times 1 \times 10^{-3}\ m^3 = 1.8\ m^3$
   d) $0.1\ dm^3 = 0.1 \times 1 \times 10^{-3}\ m^3 = 1 \times 10^{-4}\ m^3$

3 Algebra

1. a) $V = \frac{nRT}{p}$

   b) $n = \frac{pV}{RT}$

   c) $T = \frac{pV}{nR}$

   d) $p = \frac{nRT}{V}$

2. $f = \frac{E}{h}$

3. $m = \frac{(y - c)}{x}$

4. a) $\lambda = \frac{c}{f}$

   b) $V = \frac{n}{c}$

   c) $\Delta T = \frac{Q}{mc}$

   d) $V = \frac{1000n}{c}$

   e) $c = \frac{1000n}{V}$

   f) $h = \frac{E}{f}$

   g) $\Delta S = \frac{(\Delta H - \Delta G)}{T}$

4 Graphs

1. a) 0.42
   b) 2.00
   c) −8.33
   d) −0.03
   e) 80.00
   f) 0.60

2. a) $y = -0.033x + 0.6$

   b) $y = 0.375x + 10$ or $y = \frac{3}{8}x + 10$

   c) $y = 50x$
   d) $y = 2x + 2$
   e) $y = 3.5x - 2$
   f) $y = -20x - 1$

5 Geometry

1. $\frac{360}{3} = 120°$

2. a)

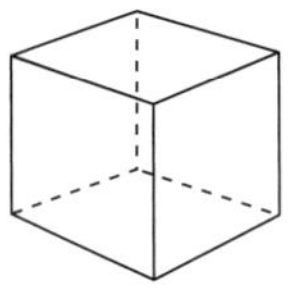

   b)

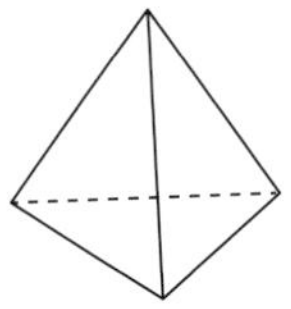

   c)

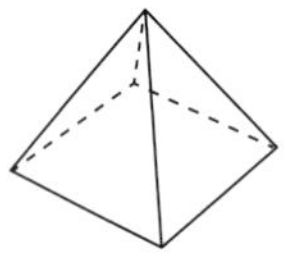

   d)

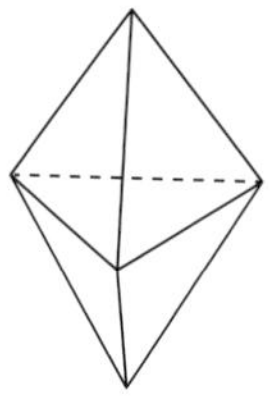

   e)

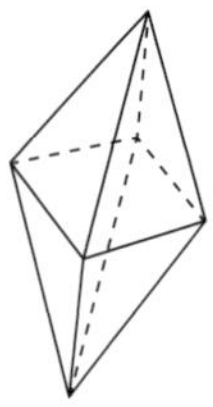

3. D

4. B

6 The amount of a substance

1. a) Carbon monoxide (CO) $M_r = 12.0 + 16.0 = 28$
   b) Phosphorus pentachloride (PCl_5)
   $M_r = 31 + (5 \times 35.5) = 208.5$
   c) Propane (C_3H_8) $M_r = (3 \times 12.0) + (8 \times 1.0) = 44$

d) Ethanoic acid (CH_3COOH)
$M_r = 2 \times 12.0 + 2 \times 16.0 + 4 \times 1.0 = 60$

e) Ethanol (C_2H_5OH)
$M_r = 2 \times 12.0 + 16.0 + 6 \times 1.0 = 46$

2 a) Sodium sulfate (Na_2SO_4)
Relative formula mass = $2 \times 23.0 + 32.1 + 4 \times 16.0$
= 142.1

b) Potassium hydroxide (KOH)
Relative formula mass = 39.1 + 16.0 + 1.0 = 56.1

c) Hydrochloric acid (HCl)
Relative formula mass = 1 + 35.5 = 36.5

d) Sodium hydroxide (NaOH)
Relative formula mass = 23.0 + 16.0 + 1 = 40.0

e) Potassium iodide (KI)
Relative formula mass = 39.1 + 126.9 = 166.0

f) Sodium carbonate (Na_2CO_3)
Relative formula mass = $2 \times 23 + 12.0 + 3 \times 16 = 106$

3 In $2CO_2$ there are 2 moles of C atoms and 4 moles of O atoms, which is a total of 6 moles of atoms. Each mole contains 6.02×10^{23} atoms.
Hence number of atoms = $6 \times 6.02 \times 10^{23}$
= 3.612×10^{24} atoms

4 a) $C_4H_{10} + 6.5O_2 \longrightarrow 4CO_2 + 5H_2O$
$2C_4H_{10} + 13O_2 \longrightarrow 8CO_2 + 10H_2O$
Balance the carbon and the hydrogen first. Leave balancing the oxygen till last as it appears in two of the products. Use halves to balance and then double up at the end.

b) Rearranging $n = \frac{m}{M}$ for m gives $m = Mn$

$m = 58 \times 5 = 290$ g
The M of butane is 58 g mol^{-1}.

5 In 0.0150 mol of S_8 molecules there would be $0.0150 \times 8 = 0.12$ mol of S atoms
In 1 mol of S atoms there are 6.02×10^{23} atoms
Hence, in 0.12 mol of S atoms there are
$0.12 \times 6.02 \times 10^{23} = 7.224 \times 10^{22}$ atoms.

6 Mr of $Na_2S_2O_3 = (2 \times 23.0) + (2 \times 32.1) + (3 \times 16.0)$
= 158.2
There are two moles of S in one mole of $Na_2S_2O_3$.
2 moles of S have a mass of $2 \times 32.1 = 64.2$
Percentage by mass of sulfur

$$= \frac{\text{mass of sulfur}}{\text{mass of sodium thiosulfate}} \times 100$$

$$= \frac{64.2}{158.2} \times 100$$

= 40.6% (3 s.f.)

7 Number of moles of iodine molecules in 0.025 g,

$$n = \frac{m}{M} = \frac{0.025}{(2 \times 127)}$$

$$= \frac{0.025}{254}$$

$= 9.84 \times 10^{-5}$ mol
Number of moles of iodine atoms in 0.025 g,
$n = 2 \times 9.84 \times 10^{-5} = 1.97 \times 10^{-4}$
Number of iodine atoms in
0.025 g = $n \times L$
$= 1.97 \times 10^{-4} \times 6.022 \times 10^{23}$
$= 1.19 \times 10^{20}$ (3 s.f.)

8 a) Relative atomic mass,

$$Ar = \frac{\text{Average mass of one atom of an element}}{\frac{1}{12}\text{Mass of one atom of carbon–12}}$$

b) i) $n = \frac{m}{A_r} = \frac{234}{118.7} = 1.9714 = 1.97$ mol (3 s.f.)

ii) Number of atoms in
n moles = $n \times L$ or $n \times N_A$
$= 1.9714 \times 6.02 \times 10^{23}$
$= 1.1868 \times 10^{24} = 1.19 \times 10^{24}$ (3 s.f.)

7 Balancing and using chemical equations to predict masses

1 a) $Mg + 2HCl \longrightarrow MgCl_2 + H_2$
b) $2Na + 2H_2O \longrightarrow 2NaOH + H_2$
c) $N_2 + 3H_2 \longrightarrow 2NH_3$
d) $Cl_2 + 2KBr \longrightarrow 2KCl + Br_2$
e) $4C + 5H_2 \longrightarrow C_4H_{10}$
f) $2PbS + 3O_2 \longrightarrow 2PbO + 2SO_2$
g) $BCl_3 + 3H_2O \longrightarrow H_3BO_3 + 3HCl$
h) $CaCO_3 + 2HCl \longrightarrow CaCl_2 + CO_2 + H_2O$
i) $2NaOH + CO_2 \longrightarrow Na_2CO_3 + H_2O$
j) $2NaOH + H_2SO_4 \longrightarrow Na_2SO_4 + 2H_2O$

2 a) $Ba(OH)_2 + H_2SO_4 \longrightarrow BaSO_4 + 2H_2O$
b) $Ba(OH)_2 + H_2SO_4 \longrightarrow BaSO_4 + 2H_2O$
1 mole 1 mole
The M_r of $Ba(OH)_2$ and $BaSO_4$ are found.
171.3 g 233.4
$\frac{171.3}{171.3} = 1$ g $\frac{233.4}{171.3}$ g
2.0 g $\frac{233.4}{171.3} \times 2 = 2.72$ g

3 a) i) Molar mass of $NH_3 = (1 \times 14.0) + (3 \times 1.01)$
= 17.03 g mol^{-1}
The molar mass and relative molecular mass are the same numerically. The units are different.

ii) Molar mass of
$NH_4NO_3 = (2 \times 14.0) + (4 \times 1.01) + (3 \times 16.0)$
$= 80.04\ g\ mol^{-1}$

b) Now according to the equation one mole of ammonia reacts with one mole of nitric acid to produce one mole of ammonium nitrate.
Hence 17.03 g of NH_3 (i.e. the molar mass) react to give 80.04 g of NH_4NO_3 (i.e. the molar mass).
Now we change these amounts to tonnes so 17.03 tonnes of NH_3 react to give 80.04 tonnes of NH_4NO_3
Now the amount of NH_3 is 34.06 tonnes so we double the amounts above
Hence 34.06 tonnes of NH_3 would produce 160.08 tonnes of NH_4NO_3

4 a) i) M for ammonia $= (1 \times 14.00) + (3 \times 1.01)$
$= 17.03\ g\ mol^{-1}$

ii) M for ammonium sulfate
$= (2 \times 14.00) + (8 \times 1.01) + (1 \times 32.07) + (4 \times 16.00)$
$= 132.15\ g\ mol^{-1}$

b) 17.03 g of ammonia produces 132.15 g of ammonium sulfate

$\frac{17.03}{17.03} = 1$ g of ammonia produces

$\frac{132.15}{17.03} = 7.76$ g of ammonium sulfate

1 tonne of ammonia produces 7.76 tonnes of ammonium sulfate
20 tonnes of ammonia produces 20 × 7.76 = 155.2 tonnes of ammonium sulfate

5 Number of moles of ethanol in 230 g $= \frac{230}{46} = 5$

M for ethanol is $46\ g\ mol^{-1}$.
Now the number of moles of glucose needed is half this according to the stoichiometry of the equation. Hence,

$$C_6H_{12}O_6 \longrightarrow 2C_2H_5OH + 2CO_2$$
2.5 moles 5 moles

Using $n = \frac{m}{M}$

The mass, m, in 2.5 moles of glucose is found. M for glucose is $180\ g\ mol^{-1}$.
Rearranging gives $m = nM = 2.5 \times 180 = 450$ g

8 Radioactive decay and half-life

1 Number of half-lives in 30.18 hours $= \frac{30.18}{10.6} = 3$

Amount of ^{212}Pb remaining $= \frac{1}{2} \times \frac{1}{2} \times \frac{1}{2} \times 56$ mg = 7 mg

2 Number of half-lives in 11 460 years $= \frac{11\,460}{5730} = 2$

Fraction of ^{14}C remaining $= \frac{1}{2} \times \frac{1}{2} = \frac{1}{4}$

3 Number of half-lives in 90 years $= \frac{90}{30} = 3$

Amount of ^{137}Cs remaining $= \frac{1}{2} \times \frac{1}{2} \times \frac{1}{2} \times 100$ g = 12.5 g

9 Percentage yield and atom economy

1 % atom economy $= \frac{\text{molar mass of desired product}}{\text{molar mass of all the products}} \times 100$

$= \frac{112}{(112 + 132)} \times 100$
$= 45.9016$
$= 45.9\%$

The answer is given correct to 3 s.f. as all the M values used in the calculation were given to 3 s.f.

2 a) Number of moles in 70 g of ethene $= \frac{70}{28} = 2.5$

To calculate the number of moles, the equation $n = \frac{m}{M}$ is used.

According to the equation, one mole of ethene produces one mole of chloroethane so 2.5 moles of ethene produce 2.5 moles of chloroethane.
One mole of chloroethane has a mass of 64.5 g, so 2.5 moles would have a mass of 2.5 × 64.5 = 161.25 g.

You could use the equation $n = \frac{m}{M}$ and make m the subject of the equation by multiplying both sides of the equation by M to give $m = nM$. The values are then substituted in to give m, the required mass.

Max theoretical mass of chloroethane = 161 g (3 s.f.)

b) Percentage yield $= \frac{\text{Actual yield}}{\text{Theoretical yield}} \times 100$

$= \frac{119}{161} \times 100$
$= 73.9\%$ (3 s.f.)

c) % atom economy $= \frac{\text{mass of useful product}}{\text{mass of all the products}} \times 100$

$= \frac{64.5}{64.5} \times 100$
$= 100\%$

When there is only one product produced, the % atom economy is always 100%.

3 From the equation 28.0 g of ethene theoretically produces 46.0 g of ethanol.

Hence 28.0 tonnes of ethene theoretically produces 46.0 tonnes of ethanol.

$$\text{percentage yield} = \frac{\text{Actual yield}}{\text{Theoretical yield}} \times 100$$

$$= \frac{43.7}{46.0} \times 100 = 95\%$$

Notice the way the units can be changed from g to tonnes.

10 Determining relative atomic mass from relative or percentage abundance

1. Relative atomic mass, $A_r = \left(\frac{18.70}{100} \times 10\right) + \left(\frac{81.30}{100} \times 11\right)$

$= 10.81$ (4 s.f.)

Remember to check your answer by asking yourself if the answer is reasonable. Clearly the answer has to be between 10 and 11 and it is much nearer 11 because of the high % abundance of the heavier isotope.

2. $A_r = \left(\frac{72.5}{100} \times 63\right) + \left(\frac{27.5}{100} \times 65\right) = 63.55 = 63.6$ (3 s.f.)

3. $A_r = \frac{(82 \times 3.0) + (83 \times 2.0) + (84 \times 10.0) + (86 \times 3.0)}{(3.0 + 2.0 + 10.0 + 3.0)}$

$= 83.9$ (1 d.p.)

11 Concentrations

1. $50\ cm^3 = \frac{50}{1000}\ dm^3 = 0.05\ dm^3$

You need to spot in the question that $0.500\ mol\ dm^{-3}$ is a concentration.

Now, $c = \frac{n}{V}$

Rearranging gives $n = cV$

Hence $n = 0.500 \times 0.05 = 0.025$ moles

2. Number of moles of NaOH $= \frac{m}{M}$

$= \frac{10}{(23.0 + 16.0 + 1.0)}$

$= 0.25$ mol

Volume of solution $= 400\ cm^3$

$= \frac{400}{1000}\ dm^3 = 0.4\ dm^3$

Using $c = \frac{n}{V}$ gives $c = \frac{0.25}{0.4} = 0.625\ mol\ dm^{-3}$

3. M_r of ethanol (C_2H_5OH) $= (2 \times 12.0) + (1 \times 16.0) + (6 \times 1)$

$= 46\ g\ mol^{-1}$

Number of moles of ethanol in 80 mg $= \frac{m}{M}$

$= \frac{80 \times 10^{-3}}{46}$

$= 1.7391 \times 10^{-3}$ mol

$100\ cm^3 = \frac{100}{1000}\ dm^3 = 0.1\ dm^3$

Now in $1\ dm^3$ there would be ten times the number of moles in $0.1\ dm^3$.

Hence concentration $= 10 \times 1.7391 \times 10^{-3}$

$= 1.74 \times 10^{-2}\ mol\ dm^{-3}$.

12 The ideal gas equation

1. Using $pV = nRT$ and rearranging for n gives

$$n = \frac{pV}{RT} = \frac{101 \times 10^3 \times 638 \times 1 \times 10^{-6}}{8.31 \times 298} = 2.60 \times 10^{-2}\ mol$$

2. Using $pV = nRT$ and rearranging for V gives

$$V = \frac{nRT}{p} = \frac{2.5 \times 10^{-2} \times 8.31 \times 323}{150 \times 10^3} = 4.47 \times 10^{-4}\ m^3$$

3. Using $pV = nRT$ and rearranging for p gives

$$p = \frac{nRT}{V} = \frac{10 \times 8.31 \times 298}{2000 \times 1 \times 10^{-6}}$$

$= 123\,819\,000$ Pa $= 123\,819$ kPa

Hence, pressure of gas $= 124000$ kPa (3 s.f.)

4. a) Potassium nitrate is KNO_3

$M = 39.1 + 14.0 + (3 \times 16.0) = 101.1\ g\ mol^{-1}$

Number of moles, $n = \frac{m}{M} = \frac{1.00}{101.1} = 9.89 \times 10^{-3}$ mol

b) Using $pV = nRT$ and rearranging for n gives

$$n = \frac{pV}{RT} = \frac{100 \times 10^3 \times 1.22 \times 10^{-4}}{8.31 \times 298} = 4.93 \times 10^{-3}\ mol$$

13 Acid-base titrations

1. $2KOH(aq) + H_2SO_4(aq) \longrightarrow K_2SO_4(aq) + 2H_2O(l)$

2 moles 1 mole

No of moles of H_2SO_4 in $45\ cm^3$ of $0.100\ mol\ dm^{-3}$,

$n = cV = 0.100 \times \frac{45}{1000} = 0.0045$ mol

0.0045 moles of H_2SO_4 require

$2 \times 0.0045 = 0.009$ moles of KOH

Concentration of KOH $= \frac{n}{V} = \frac{0.009}{0.03} = 0.3\ mol\ dm^{-3}$

2. a) $MgO + 2HCl \longrightarrow MgCl_2 + H_2O$

1 mole 2 moles

0.02 moles 0.04 moles

MgO is a basic oxide. It reacts with the acid to give a salt (i.e. $MgCl_2$) and water.
The ratio of MgO to HCl is 1:2.
Hence 0.04 moles of acid are needed.

b) Volume of acid needed = $30\ cm^3 = \frac{30}{1000}\ dm^3$

$= 0.030\ dm^3$

Using $c = \frac{n}{V} = \frac{0.04}{0.030} = 1.33\ mol\ dm^{-3}$ (3 s.f.)

3 $2NaOH(aq) + H_2SO_4(aq) \longrightarrow Na2SO_4(aq) + 2H_2O(l)$
2 moles 1 mole
No of moles of H_2SO_4 in 25 cm³ of 0.10 mol dm⁻³,

$n = cV = 0.10 \times \frac{25}{1000} = 0.0025$ mol

0.0025 moles of H_2SO_4 require
$2 \times 0.0025 = 0.005$ moles of NaOH

Concentration of NaOH, $c = \frac{n}{V}$. Rearranging this equation for V, gives

$V = \frac{n}{c} = \frac{0.005}{0.08} = 0.0625\ dm^3$

$= 0.0625 \times 1000\ cm^3 = 62.5\ cm^3$

14 Determining empirical and molecular formulae

1 a) C H

$\frac{92.3}{12.0} = 7.6917 \quad \frac{7.7}{1.0} = 7.7$

$\frac{7.6917}{7.6917} \approx 1 \quad \frac{7.7}{7.6917} \approx 1$

Empirical formula is CH

b) The empirical formula mass $= (1 \times 12.0) + (1 \times 1.0)$
$= 13.0$

The relative molecular mass = 78.06
= 6 × empirical formula mass.

Hence molecular formula is C_6H_6

2 %O = 100 − (25.5 + 36.5) = 38%

S Na O

$\frac{25.5}{32.1} = 0.7944 \quad \frac{36.5}{23.0} = 1.5870 \quad \frac{38}{16.0} = 2.3750$

$\frac{0.7944}{0.7944} = 1 \quad \frac{1.5870}{0.7944} \approx 2 \quad \frac{2.3750}{0.7944} \approx 3$

Empirical formula is Na_2SO_3

3 a) C O Cl

$\frac{12.1}{12.0} \quad \frac{16.2}{16.0} \quad \frac{71.7}{35.5}$

$= 1.01 \quad = 1.01 \quad = 2.02$

$\frac{1.01}{1.01} = 1 \quad \frac{1.01}{1.01} = 1 \quad \frac{2.02}{1.01} = 2$

Empirical formula is $COCl_2$

b) You would need the M_r of phosgene.

4 72.2% by mass of magnesium means there must be 100 − 72.2 = 27.8% by mass of nitrogen.

Hence Mg N

$\frac{72.2}{24.3} \quad \frac{27.8}{14}$

$= 2.97 \quad = 1.99$

$\approx 3 \quad \approx 2$

Empirical formula for magnesium nitride is Mg_3N_2

15 Calculating the formula of a hydrated salt

1 Anhydrous copper(II) sulfate is $CuSO_4$
If this is taken from the formula $CuSO_9H_{10}$ we are left with $H_{10}O_5$ which is the water of crystallisation. Now $H_{10}O_5 = 5H_2O$.
Hence the formula of hydrated copper(II) sulfate is $CuSO_4.\ 5H_2O$

2 a) It shows the number of water molecules that make up the water of crystallisation.

b) Molar mass of $Na_2SO_4 = ((2 \times 23.0) + 32.1 + (4 \times 16.0))$
$= 142.1\ g\ mol^{-1}$

Molar mass of x water molecules = 322.1 − 142.1
$= 180\ g\ mol^{-1}$

Molar mass of one water molecule = (2 × 1.0 + 16.0)
$= 18\ g\ mol^{-1}$

Hence, $x = \frac{180}{18} = 10$

3 a) Relative formula mass of $Na_2S_2O_3.5H_2O$
$= ((2 \times 23.0) + (2 \times 32.1) + (3 \times 16.0) + (5 \times 18))$
$= 248.2$

Note that relative formula mass does not have any units.

b) No of moles in 25 g of $Na_2S_2O_3.5H_2O = \frac{25}{248.2}$
$= 0.1007$

Hence there would be 5 × 0.1007 = 0.5035 moles of water

1 mole of water is 18.0 g so 0.5035 moles would have a mass of 0.5035 × 18 = 9.06 g (3 s.f.)

16 Enthalpy changes and calorimetry

1 Heat given to the water, $Q = mc\Delta T$
$= 100.0 \times 4.2 \times 1.8$
$= 756\ J$

To convert from J to kJ you divide by 1000.

= 0.756 kJ

Relative formula mass of $CuSO_4$ = 63.5 + 32.1 + 4 × 16.0

= 159.6 g mol^{-1}

Number of moles, $n = \frac{m}{M} = \frac{12}{159.6} = 0.07519$ mol

Enthalpy change, $\Delta H = -\frac{0.756}{0.07519} = -10.1$ kJ mol^{-1}

Remember to insert the minus sign as this is an exothermic reaction.

2 We can create a diagram putting the arrows in the direction from reactants to products and reversing any enthalpy changes where necessary.

$H^+(g) + Cl^-(g) \xrightarrow{\Delta H_r^\ominus} H^+(aq) + Cl^-(aq)$

$H^+(g) + Cl^-(g)$ ↓ −963 → $H(g) + Cl(g)$

$H(g) + Cl(g) \xrightarrow{-432} HCl(g)$

$HCl(g)$ ↑ −75 → $H^+(aq) + Cl^-(aq)$

Don't waste time and space drawing diagrams again. Instead use the diagram in the question and alter it (i.e. changing direction of arrows and putting values in).

$\Delta H_r = (-963) + (-432) + (-75)$

$= -1470$ kJ mol^{-1}

Hess's law is used. Here an alternative route from reactants to products is shown in the diagram. We then calculate the enthalpy changes along this path.

3 a) Standard enthalpy change of formation for $O_2(g) = 0$ kJ mol^{-1} (because it is already an element in its standard state).

b)

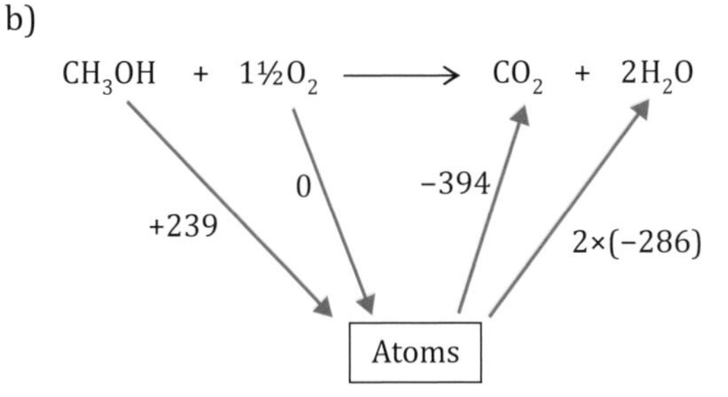

$\Delta H_c^\ominus = +239 + 0 + (-394) + 2 \times (-286)$

$= -727$ kJ mol^{-1}

4 Standard enthalpy of solution = standard lattice enthalpy of dissociation of magnesium chloride + standard enthalpy of hydration of $Mg^{2+}(g)$ + (2 × standard enthalpy of hydration of $Cl^-(g)$)

Hence, substituting the values in, we have:

−155 = 2493 + (−1920) + (2 × standard enthalpy of hydration of $Cl^-(g)$)

−728 = (2 × standard enthalpy of hydration of $Cl^-(g)$)

Hence, standard enthalpy of hydration of

$Cl^-(g) = \frac{-728}{2} = -364$ kJ mol^{-1}

5 $-288 = 417 + \Delta H_L^\ominus$

Solving gives $\Delta H_L^\ominus = -705$ kJ mol^{-1}

6 a) A = Standard enthalpy change of formation of magnesium chloride

B = Standard enthalpy change of atomisation of magnesium

C = Standard second ionisation energy of magnesium

b) E = 2 × $\Delta H_{at}^\ominus$ = 2 × (+121) = +242 kJ mol^{-1}

c) Using Hess's law we have:

$-642 = 150 + 736 + 1450 + 242 - 728 + \Delta H_L^\ominus$

Solving, gives $\Delta H_L^\ominus = -2492$ kJ mol^{-1}

17 Bond enthalpies

1

CH_4 (H—C—H with two further H on C) + O=O, O=O ⟶ O=C=O + H—O—H, H—O—H

Enthalpy change when bonds are broken:

4 × C—H + 2 × O=O = (4 × 413) + (2 × 497) = 2646 kJ mol^{-1}

Enthalpy change when bonds are formed:

2 × C=O + 4 × O—H = (2 × 805)+(4 × 463)

= 3462 kJ mol^{-1}

Enthalpy change for reaction = Enthalpy change in breaking bonds − Enthalpy change in forming bonds

= 2646 − 3462 = −816 kJ mol^{-1}

2

H—C(H)(H)—C(H)(H)—H + Cl—Cl ⟶ H—C(H)(H)—C(H)(H)—Cl + H—Cl

Enthalpy change when bonds are broken:

6 × C—H + 1 × C—C + 1 × Cl—Cl

= (6 × 413) + (1 × 347) + (1 × 243)

= 3068 kJ mol^{-1}

Enthalpy change when bonds are formed:

5 × C—H + 1 × C—C + 1 × C—Cl + 1 × H—Cl

= (5 × 413) + (1 × 347) + (1 × 346) + (1 × 432)

= 3190 kJ mol^{-1}

Enthalpy change for reaction = Enthalpy change in breaking bonds − Enthalpy change in forming bonds = 3068 − 3190 = −122 kJ mol^{-1}

3 a) $\frac{1}{2}N_2(g) + \frac{3}{2}H_2(g) \longrightarrow NH_3(g)$

b) Enthalpy change when bonds are broken:
$\frac{1}{2} \times N{\equiv}N + \frac{3}{2} \times H{-}H = (\frac{1}{2} \times 944) + (\frac{3}{2} \times 436)$
$= 1126\ kJ\ mol^{-1}$
Enthalpy change when bonds are formed:
$3 \times N{-}H = (3 \times 388) = 1164\ kJ\ mol^{-1}$
Enthalpy change for reaction = Enthalpy change in breaking bonds − Enthalpy change in forming bonds = 1126 − 1164 = −38 kJ mol^{-1}

4 Enthalpy change when bonds are broken:
$= 4 \times N{-}H + 1 \times N{-}N + 4 \times O{-}H + 2 \times O{-}O$
$= (4 \times 388) + (1 \times 163) + (4 \times 463) + (2 \times 146)$
$= 3859\ kJ\ mol^{-1}$
Enthalpy change when bonds are formed:
$1 \times N{\equiv}N + 8 \times O{-}H = (1 \times 944) + (8 \times 463)$
$= 4648\ kJ\ mol^{-1}$
Enthalpy change for reaction = Enthalpy change in breaking bonds − Enthalpy change in forming bonds = 3859 − 4648 = −789 kJ mol^{-1}

18 Rates of reaction

1 a) 1
b) 3
c) 3 + 1 = 4
d) i) It would double
ii) It would increase by a factor of $2^3 = 8$

2 rate = k[A][B]
$[A]^1 = [A]$ and $[B]^1 = [B]$.

3 a) From experiments 1 and 2, [A] is doubled and as this is squared in the rate equation, the initial rate will go up by a factor of 4. Hence, initial rate for experiment 2 = $4 \times 4 \times 10^{-5} = 1.6 \times 10^{-4}$.
From experiments 1 and 3, [B] is doubled. As the rate is proportional to [B] the rate would normally double, but here it has stayed the same. The concentration of [A] must have decreased. Now $[A]^2$ must have halved. This means that

the concentration must be $\frac{[A]}{\sqrt{2}}$

(note that $\frac{[A]}{\sqrt{2}} \times \frac{[A]}{\sqrt{2}} = \frac{[A]^2}{2}$.)

Hence in experiment 3, initial concentration of

$A = \frac{1.5 \times 10^{-2}}{\sqrt{2}} = 1.06 \times 10^{-2} = 1.1 \times 10^{-2}$ (2 s.f.)

From experiments 1 and 4, the initial rate has

increased by a factor of $\frac{3.6 \times 10^{-4}}{4.0 \times 10^{-5}} = 9$.

Now the initial [B] has increased by a factor of 3.
Which means $[A]^2$ has increased by a factor of 3 (to make 9 when the concentrations are multiplied together). Hence the concentration of A has increased by a factor of $\sqrt{3}$. The new concentration of A = $\sqrt{3} \times 1.5 \times 10^{-2} = 2.6 \times 10^{-2}$ (2s.f.)
The table can now be completed to give:

Experiment	Initial concentration of A/mol dm^{-3}	Initial concentration of B/mol dm^{-3}	Initial rate/ mol dm^{-3}s^{-1}
1	1.5×10^{-2}	2.1×10^{-2}	4.0×10^{-5}
2	3.0×10^{-2}	2.1×10^{-2}	**1.6×10^{-4}**
3	**1.1×10^{-2}**	4.2×10^{-2}	4.0×10^{-5}
4	**2.6×10^{-2}**	6.3×10^{-2}	3.6×10^{-4}

b) rate = $k[A]^2[B]$

Rearranging, gives $k = \frac{\text{rate}}{[A]^2[B]}$

$= \frac{4 \times 10^{-5}}{[1.5 \times 10^{-2}]^2[2.1 \times 10^{-2}]}$

= 8.5 (2 s.f.)

$$k = \frac{\text{rate}}{[A]^2[B]} = \frac{\text{mol dm}^{-3}\text{ s}^{-1}}{(\text{mol dm}^{-3})^2(\text{mol dm}^{-3})}$$

$$= \frac{\text{s}^{-1}}{(\text{mol dm}^{-3})^2}$$

$$= \frac{\text{s}^{-1}}{\text{mol}^2\text{ dm}^{-6}}$$

$$= \text{mol}^{-2}\text{ dm}^6\text{ s}^{-1}$$

4 a) rate = $k[NO]^2[H_2]$

$$k = \frac{\text{rate}}{[NO]^2[H_2]}$$

$$= \frac{7.92 \times 10^{-7}}{(1.5 \times 10^{-2})^2(1.1 \times 10^{-2})}$$

= 0.32

$$k = \frac{\text{rate}}{[NO]^2[H_2]} = \frac{\text{mol dm}^{-3}\text{ s}^{-1}}{(\text{mol dm}^{-3})^2(\text{mol dm}^{-3})}$$

As the unit mol dm^{-3} appears in the top (numerator) and the bottom (denominator) of the fraction, they can be cancelled.

$$= \frac{\text{s}^{-1}}{(\text{mol dm}^{-3})^2}$$

When the units in the bottom of the fraction are brought to the top, the sign of any powers to which they are raised is reversed.

$$= \frac{\text{s}^{-1}}{(\text{mol}^2\text{ dm}^{-6})}$$

$$= \text{mol}^{-2}\text{ dm}^6\text{ s}^{-1}$$

b) Now as rate = $k[NO]^2[H_2]$, halving $[H_2]$ would halve the rate and doubling [NO] would quadruple the rate, so overall the rate would increase by a factor of $\frac{1}{2} \times 4 = 2$.
Hence the initial rate of reaction
$= 2 \times 7.92 \times 10^{-7} = 1.584 \times 10^{-6}$ mol dm^{-3} s^{-1}.

5 a) Comparing the results of experiments 1 and 2. The initial concentration of $S_2O_8^{2-}$ has doubled from 0.0200 to 0.0400 whilst the initial concentration of I^- has stayed constant. This doubling has produced a doubling in the initial rate. The order of the reaction with respect to persulfate ions is 1.

b) Comparing the results of experiments 2 and 3. The initial concentration of I^- has doubled from 0.0100 to 0.0200 whilst the initial concentration of $S_2O_8^{2-}$ has stayed constant. This doubling has produced a doubling in the initial rate. The order of the reaction with respect to iodide ions is 1.

c) rate = $k[S_2O_8^{2-}][I^-]$
We have used the results from experiment 1 to calculate the value of k. We could have used results from experiments 2 or 3 to obtain the same answer.

$$k = \frac{\text{rate}}{[S_2O_8^{2-}][I^-]}$$

$$= \frac{4.00 \times 10^{-6}}{(0.0200)(0.0100)}$$

$$= 0.02$$

For the units of k,

$$k = \frac{\text{rate}}{[S_2O_8^{2-}][I^-]} \quad \frac{\text{mol dm}^{-3}\text{ s}^{-1}}{(\text{mol dm}^{-3})(\text{mol dm}^{-3})}$$

$$= \frac{\text{s}^{-1}}{(\text{mol dm}^{-3})} = \text{mol}^{-1}\text{ dm}^3\text{ s}^{-1}$$

19 The shapes of simple molecules

1 a) Xe is in Group 8, so there are 8 electrons plus one from each of the 4 fluorine atoms giving a total of 12 electrons. This is 6 pairs of electrons. Since there are four bonds to the fluorine, there will be 4 bonding pairs leaving 2 lone pairs.
Number of bonding pairs = 4

b) Number of lone pairs = 2

c)

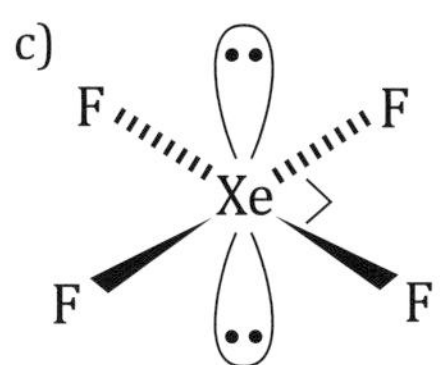

Note all the fluorine atoms are in the same plane. The two lone pairs are at 90° to this plane.

d) The shape of XeF_4 is square planar.

2 Phosphorus is in Group 5 and therefore has 5 electrons in the outer shell. The four hydrogen atoms in the ion will give another 4 electrons, making 9.
As there is a +1 charge on the ion, we need to subtract one electron. There are a total of 8 electrons resulting in 4 bonding pairs. There will be equal repulsion between the 4 bonding pairs of electrons so the ion will adopt a tetrahedral structure and the bond angle will be 109.5°.

3 For BrF_4^-
Bromine, the central atom in the ion, is in Group 7 so there are 7 outer electrons.
There are four fluorine atoms each contributing another 4 electrons and as there is a negative charge there is also another electron making a total of 12 electrons. There are 6 electron pairs with 4 of them being used in bonding to the fluorine. Hence there are two lone pairs.
The molecule will be square planar with bond angles of 90°.

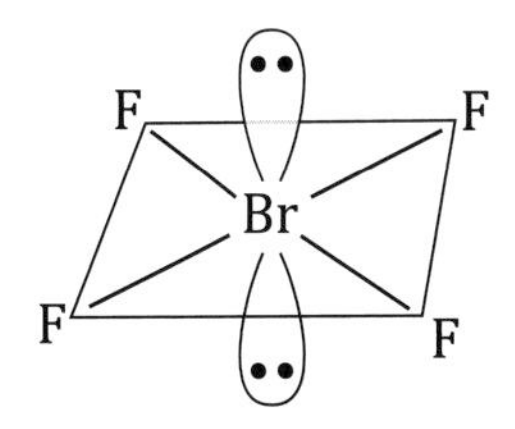

The two lone pairs will be situated above and below the plane containing all the atoms. The two lone pairs will now be as far away from each other as possible.

For BrF_2^+
Bromine, the central atom in the ion, is in Group 7 so there are 7 outer electrons.
There are two fluorine atoms each contributing another 2 electrons and as there is a positive charge, there is one fewer electron so this makes 8 electrons in total. Now 8 electrons means 4 electron pairs and since there are two bonds to fluorine, there will be two bonding pairs and two lone pairs.
The orbitals will adopt a tetrahedral arrangement and repulsion from the lone pairs will close the bond angle slightly (i.e. from 109.5° for a tetrahedral arrangement) to 104.5°. As the lone pairs do not appear in the shape, the molecule will be a bent molecule.

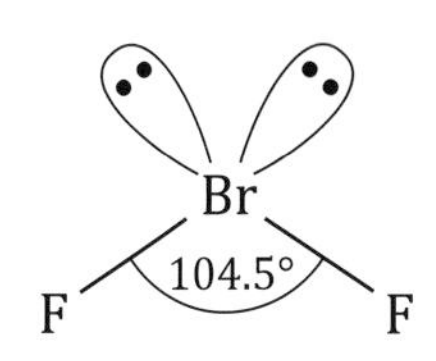

20 The solubility of compounds in water

1 a) Solubility is 8 g of solute per 100 g of solution.

b) Solubility at 70°C is 34 g of solute per 100 g of solution.
Solubility at 30°C is 12 g of solute per 100 g of solution.
Hence,
maximum mass = difference in the solubilities
= 34 − 12
= 22 g

2 At 10°C a 200 cm³ saturated solution would contain

$\frac{45.0}{5} = 9.0$ g of X

At 30°C a 200 cm³ saturated solution would contain

$\frac{64.0}{5} = 12.8$ g of X

Hence, mass produced = 12.8 − 9.0 = 3.8 g

21 Oxidation states (or numbers)

1 a) Each O has an oxidation state of −2, so as there are 2 of these, this will give a contribution of −4.
As the compound is neutral, the carbon must have an oxidation state of +4 (because − 4 + 4 = 0)

b) Each O has an oxidation state of −2, so as there are 4 of these, this will give a contribution of −8. Each potassium has an oxidation state of +1 so this gives +2. Adding these gives −8 + 2 = −6
As the compound is neutral, the sulfur must have an oxidation state of +6 (because −6 +6 = 0)

c) Oxygen (provided it is combined) has an oxidation state of −2. As there are two sodiums each must have and oxidation state of +1.

d) Magnesium has an oxidation state of +2. The three oxygens will give a total of −6. Adding these gives −4. Hence the combined carbon has an oxidation state of +4.

e) Each O has an oxidation state of −2, so as there are 3 of these, this will give a contribution of −6. The two Fe atoms combined must have a total oxidation state of +6. Dividing this by two, gives the oxygen state of Fe as +3.

2 a) +6
b) +2
c) +4
d) +3
e) +1
f) +5
g) +4
h) +7
i) +7
j) +6

3 Calcium and oxygen have oxidation states of +2 and −2 respectively.
There are 8 atoms of oxygen which give 8 × (−2) = −16.
Adding the +2 from the calcium gives −14.
The two Mn atoms must add up to +14.
Giving oxidation state of Mn = +7

4 Sodium and oxygen have oxidation states of +1 and −2 respectively. There are 3 oxygen atoms giving −6 which when added to the single sodium gives −5. Hence the oxidation state of the Cl is +5.

5 $2Na \longrightarrow 2Na+ + 2e^-$ This is oxidation as the sodium loses electrons.
$Cl_2 + 2e^- \longrightarrow 2Cl^-$ This is reduction as the chlorine gains electrons.

22 Redox reactions

1 a) Redox reactions are those reactions where the atoms have their oxidation numbers/states changed. One atom gains electrons whilst the other loses electrons.

b) NO – Combined oxygen has an oxidation state of −2. Hence the N will have an oxidation state of +2.
NH_4^+ – Combined hydrogen has an oxygen state of +1 so as there are 4 such atoms this gives +4. Now the oxidation state of N added to +4 needs to give the overall charge on the ion, which is +1. Hence the oxidation state of N is −3.
Now as the oxidation state of N has decreased (i.e. gone from +2 to −3), the nitrogen has been reduced.
Remember an increase in oxidation state represents oxidation and a decrease represents reduction.

2 a) The oxidising agent gains electrons so it is reduced.

b) i) The HCl(aq)
ii) Zn(s) oxidation state is zero as it is an uncombined element.
Zn in $ZnCl_2$ has an oxidation state of +2 (as each chlorine has −1 and there are two of them).
iii) The oxidation state increases (i.e. from 0 to +2) so the Zn is oxidised.

3 a)

Element	Initial oxidation state	Final oxidation state
nitrogen	−3	2
hydrogen	1	1
oxygen	0	−2

b) From the table, the nitrogen has increased its oxidation state (i.e. from −3 to 2).
Hence the nitrogen has been oxidised.

23 Electrochemical cells

1 $E^\ominus_{cell} = E^\ominus_{right\text{-}hand\ electrode} - E^\ominus_{left\text{-}hand\ electrode}$
$2.15 = 1.69 - E^\ominus_{left\text{-}hand\ electrode}$
Rearranging, gives $E^\ominus_{left\text{-}hand\ electrode} = 1.69 - 2.15$
$= -0.46$ V

2 a) The two relevant half-equations obtained from the table are:
$Al^{3+}(aq) + 3e^- \longrightarrow Al(s)$ $E^\ominus = -1.66$ V
$Fe^{2+}(aq) + 2e^- \longrightarrow Fe(s)$ $E^\ominus = -0.44$ V
The second reaction has the more positive value of $E^\ominus$ so it will be the right-hand electrode.
$E^\ominus_{cell} = E^\ominus_{right\text{-}hand\ electrode} - E^\ominus_{left\text{-}hand\ electrode}$
$= -0.44 - (-1.66)$
$= 1.22$ V

b) The electrode with the more positive value of $E^\ominus$ is the positive electrode.
Hence, Fe is the positive electrode.

3 Obtaining the data from the table we have:
$Cl_2(g) + 2e^- \longrightarrow 2Cl^-(aq)$ $E^\ominus = +1.36$ V
$Fe^{3+}(aq) + e^- \longrightarrow Fe^{2+}(aq)$ $E^\ominus = +0.77$ V
$E^\ominus_{cell} = E^\ominus_{right\text{-}hand\ electrode} - E^\ominus_{left\text{-}hand\ electrode}$
$= 1.36 - 0.77$
$= +0.59$ V

4 a) The strongest reducing agent is the species that loses electrons most easily and this will be the species with the most negative standard electrode potential. Hence, Al is the strongest reducing agent.

b) $E^\ominus_{cell} = E^\ominus_{right\text{-}hand\ electrode} - E^\ominus_{left\text{-}hand\ electrode}$
$= 2.87 - 1.23 = +1.64$ V
The emf is positive because the standard electrode potential of F_2/F^- is higher (i.e. more positive) than the standard electrode potential of O_2/H_2O. This means that the fluorine is a more powerful oxidising agent than oxygen so it oxidises water.
The half-equation is:
$$2F_2 + 2H_2O \longrightarrow 4F^- + 4H^+ + O_2$$

5 We can start off by assuming the reaction is feasible. The standard electrode potential of chlorine (i.e. +1.36 V) is higher than that of the iron (i.e. +0.77 V), so the electrons will flow from the $Fe^{2+}(aq)$ to the $Cl_2(g)$. The more positive electrode (i.e. the chlorine) will be the right-hand electrode. The shorthand for this cell is
$Fe^{2+}(aq) | Fe^{3+}(aq) || Cl_2(g) | Cl^-(aq)$
Looking at this shorthand you can tell the following. The Fe^{2+} is oxidised to form Fe^{3+} ions, whilst the Cl_2 is reduced to Cl^- ions.
$E^\ominus_{cell} = E^\ominus_{right\text{-}hand\ electrode} - E^\ominus_{left\text{-}hand\ electrode}$
$= 1.36 - (+0.77)$
$= +0.59$ V
As this is a positive value the assumption we made is true and the reaction is feasible.

24 Entropy and free energy change

1 $\Delta S = S_{products} - S_{reactants}$
$= 189 - (131 + 0.5 \times 205)$
$= -44.5\ J\ K^{-1}\ mol^{-1}$
Note the need to multiply the S value for O_2 by half because there is only half a mole in the equation.
On the point of reaction feasibility
$\Delta G = 0$
$\Delta H = -242 \times 1000\ J\ mol^{-1}$
Using $\Delta G = \Delta H - T\Delta S$, we obtain
$0 = -242 \times 1000 - T \times (-44.5)$
Rearranging and solving, gives T = 5438 K

2 a) $S^\ominus_{reactants} = (4 \times 193) + (5 \times 205)$
$= 1797\ J\ K^{-1}\ mol^{-1}$

b) $S^\ominus_{products} = (4 \times 211) + (6 \times 189)$
$= 1978\ J\ K^{-1}\ mol^{-1}$

c) $\Delta S^\ominus = S^\ominus_{products} - S^\ominus_{reactants}$
$= 1978 - 1797$
$= +181\ J\ K^{-1}\ mol^{-1}$

3 a) $\Delta G = \Delta H - T\Delta S$
$= -727\,000 - 298 \times (-81)$
$= -702\,862\ J\ mol^{-1}$
$= -703\ kJ\ mol^{-1}$

b) The negative sign of ΔG shows that the reaction is feasible.

25 Equilibria

1 a) $K_c = \frac{[PCl_3][Cl_2]}{[PCl_5]}$

b) Units of $K_c = \frac{(mol\ dm^{-3})(mol\ dm^{-3})}{(mol\ dm^{-3})} = mol\ dm^{-3}$

2 $C_2H_5OH(l) + C_2H_5COOH(l) \rightleftharpoons C_2H_5COOC_2H_5(l) + H_2O(l)$

At equilibrium 0.07 mol $\quad$ (1 − 0.07)

4 − 0.93 = 3.07 mol $\quad$ = 0.93 mole $\quad$ 0.93 mole

There are no units for K_c as the stoichiometry of the equation means the concentration units on the top of the expression for K_c cancel with those on the bottom.

$$K_c = \frac{[C_2H_5COOC_2H_5][H_2O]}{[C_2H_5OH][C_2H_5COOH]} = \frac{(0.93)(0.93)}{(3.07)(0.07)} = 4.02$$

3 a) $K_c = \frac{[NO]^2}{[N_2O_4]}$

b) Units of $K_c = \frac{[\text{mol dm}^{-3}]^2}{[\text{mol dm}^{-3}]} = \text{mol dm}^{-3}$

Note that as the volume of the container is 1 dm^3, the concentration of both gases will be

$\frac{6}{1} = 6 \text{ mol dm}^{-3}$.

c) $K_c = \frac{[NO]^2}{[N_2O_4]} = \frac{6^2}{6} = 6$

4 a) $K_c = \frac{[NH_3]^2}{[N_2][H_2]^3}$

b) Units for $K_c = \frac{[\text{mol dm}^{-3}]^2}{[\text{mol dm}^{-3}][\text{mol dm}^{-3}]^3}$

$= \frac{[\text{mol dm}^{-3}]^2}{[\text{mol dm}^{-3}]^4}$

$= \frac{1}{[\text{mol dm}^{-3}]^2}$

$= \frac{1}{\text{mol}^2\text{ dm}^{-6}}$

$= \text{mol}^{-2}\text{ dm}^6$

c) $K_c = \frac{[\frac{NH_3}{V}]^2}{[\frac{N_2}{V}][\frac{H_2}{V}]^3} = \frac{[\frac{NH_3}{6}]^2}{[\frac{7.2}{6}][\frac{12}{6}]^3}$

Hence, $[\frac{NH_3}{6}]^2 = K_c[\frac{7.2}{6}][\frac{12}{6}]^3 = 0.08 \times 1.2 \times 8 = 0.768$

$[NH_3] = \sqrt{0.768} = 0.876 \text{ mol dm}^{-3}$

26 Calculations involving pH

1 $pH = -\log_{10}[H^+] = -\log_{10}[0.015] = 1.82$ (3 s.f.)

Note that no indication as to the numbers of significant figures or decimal places is given in the question. As the concentration is given to 3 s.f. we will give the answer to 3 s.f.

2 We first find the concentration of the acid in mol dm^{-3}.

Number of moles of HCl in 25.0 cm^3, $n = \frac{V}{1000} \times c$

$= \frac{25}{1000} \times 0.165$

$= 4.125 \times 10^{-3}$

When added to 975 cm^3 of water there will be 4.125×10^{-3} moles in (25.0 + 975) = 1000 cm^3 of aqueous solution.

The equation for the production of hydrogen ions is:

$HCl(aq) \longrightarrow H^+(aq) + Cl^-(aq)$

The number of moles of H^+ ions is the same as the number of moles of HCl. This is because HCl is a strong acid and is therefore completely dissociated into ions.

Hence, $[H^+] = \frac{n}{V} = \frac{(4.125 \times 10^{-3})}{1} = 4.125 \times 10^{-3} \text{ mol dm}^{-3}$

In the equation $= \frac{n}{V}$, the volume V must be in dm^3.

The volume is 1000 cm^3 which is equal to 1 dm^3.

$pH = -\log_{10}[H^+] = -\log_{10}[4.125 \times 10^{-3}] = 2.38$ (3 s.f.)

3 $pH = -\log_{10}[H^+]$ gives

$5 = -\log_{10}[H^+]$

$-5 = \log_{10}[H^+]$

Hence $10^{-5} = [H^+]$

$[H^+] = 1 \times 10^{-5} \text{ mol dm}^{-3}$

Note that 10^{-5} and 1×10^{-5} are the same.

4 First find the number of moles of HCl in 4.00 g. The M of HCl is 1.0 + 35.5 = 36.5 g.

$n = \frac{m}{M} = \frac{4.00}{36.5} = 0.1096 \text{ mol}$

$c = \frac{n}{V} = \frac{0.1096}{0.5} = 0.2192 \text{ mol dm}^{-3}$

The volume must be converted from cm^3 to dm^3. This is done by dividing the volume in cm^3 by 1000.

$pH = -\log_{10}[H^+] = -\log_{10}[0.2192] = 0.659$ (3 s.f.)

As hydrochloric acid is a strong acid it is completely dissociated. Also the ratio of HCl to H^+ ions is 1:1 so the concentration of H^+ ions will be 0.2192 mol dm^{-3}.

5 Number of moles of $Ca(OH)_2$, $n = \frac{m}{M} = \frac{4.0}{74.1} = 0.054 \text{ mol}$

Concentration, $c = \frac{1000n}{V}$

$= \frac{1000 \times 0.054}{500}$

$= 0.108 \text{ mol dm}^{-3}$ (3 s.f.)

As $Ca(OH)_2(aq)$ is a strong alkali, it is completely dissociated.

The equation for this is:

$Ca(OH)_2(aq) \longrightarrow Ca^{2+}(aq) + 2OH^-(aq)$

Concentration of OH^- ions = 2 × concentration of $Ca(OH)_2$

Hence, $[OH^-(aq)] = 2 \times 0.108 = 0.216 \text{ mol dm}^{-3}$

Now $[OH^-(aq)][H^+(aq)] = 1.00 \times 10^{-14} \text{ mol}^2\text{ dm}^{-6}$

Hence, $[0.216][H^+(aq)] = 1.00 \times 10^{-14}$

So, $[H^+(aq)] = \dfrac{1.00 \times 10^{-14}}{0.216} = 4.63 \times 10^{-14}\ \text{mol dm}^{-3}$

$pH = -\log_{10}[H^+] = -\log_{10}[4.63 \times 10^{-14}] = 13.3$ (3 s.f.)

6 a) i) $CH_3COOH(aq) \rightleftharpoons H^+(aq) + CH_3COO^-(aq)$

ii) $K_a = \dfrac{[H^+(aq)][CH_3COO^-(aq)]}{[CH_3COOH(aq)]}$

b) $K_a = \dfrac{[H^+(aq)]^2}{[HCOOH(aq)]}$

as $[H^+(aq)] = [HCOO^-(aq)]$ when in equilibrium

$1.7 \times 10^{-5} = \dfrac{[H^+(aq)]^2}{[0.0254]}$

$[H^+(aq)]^2 = 1.7 \times 10^{-5} \times 0.0254$

$= 4.318 \times 10^{-7}$

$[H^+(aq)] = \sqrt{4.318 \times 10^{-7}}$

$= 6.5711 \times 10^{-4}\ \text{mol dm}^{-3}$

$pH = -\log_{10}[H^+] = -\log_{10}[6.5711 \times 10^{-4}]$

$= 3.18$ (3 s.f.)

27 Analytical techniques

1 The carbonyl (i.e. C=O) group.

2 The spectrum has a peak in the 1620–1680 cm^{-1} range and according to the table of data, this indicates a double bond (i.e. unsaturation). Hence student A is correct.

3 Isomer A has 3 peaks. One peak comes from the carbons to which the NO_2 groups are attached another comes from the adjacent carbons to which the NO_2 groups are attached. The final peak comes from the 2 carbons that are furthest away from the NO_2 groups.
Isomer B has 4 peaks. One peak comes from the carbon situated between the two carbons with the NO_2 groups, and one peak comes from the carbons to which the NO_2 groups are attached. One peak comes from the next two carbons and one peak comes from the carbon on its own that is furthest from the carbons with the NO_2 groups attached.
Isomer C has 2 peaks. One peak comes from the carbons to which the NO_2 groups are attached. The other peak comes from the four remaining carbons which are all in the same environment.

4 All the hydrogens in this molecule are in the same environment so there is only one peak in the NMR spectrum.

```
        CH3
        |
  CH3—C—CH3
        |
        CH3
```

Compound X

There are three different types of environment for the hydrogens giving rise to the three peaks in the NMR spectrum.

```
    H  H  H  H  H
    |  |  |  |  |
 H—C—C—C—C—C—H
    |  |  |  |  |
    H  H  H  H  H
```

Type 3, Type 2, Type 1, Type 2, Type 3

Compound Y

5 For compound A.
The absorption at 1720 cm^{-1} indicates a C=O group. Only one peak in the 1H NMR spectrum indicates that all the hydrogens are in an equivalent environment. We can deduce from this that the molecule is likely to be symmetrical. Hence the structure is as follows.

```
    H  O  H
    |  ‖  |
 H—C—C—C—H
    |     |
    H     H
```

For compound B
Absorptions at wavenumbers 1645 cm^{-1} and 3300 cm^{-1}. The peak at 1645 cm^{-1} indicates a C=C (from the table this is at 1620–1680 cm^{-1}). The 3300 cm^{-1} peak indicates an O—H group in an alcohol (from the table this is at 3230–3550 cm^{-1}). We therefore need to draw the structure of a molecule with molecular formula C_3H_6O that contains an O—H and C=C group.
Hence the structure is as follows.

```
         H  H
 H       |  |
  \C=C—C—OH
 H/         |
            H
```

6 a) The protons/hydrogens on the methyl group are split into the doublet at 2.5 ppm by the lone proton/hydrogen on the adjacent carbon atom.

b) The lone proton/hydrogen on the carbon atom is split into the quartet at 5.8 ppm by the three protons/hydrogens on the adjacent carbon atom.

Specification map

Ticks indicate the mathematical requirements that are stated in the specifications.

Topic	Section	WJEC	AQA	OCR	Edexcel	CIE	CCEA	IB	Cam Pre-U
M1 Arithmetic and numerical computation									
Order when evaluating expressions	1.1	✓	✓	✓	✓	✓	✓	✓	✓
Directed numbers	1.3	✓	✓	✓	✓	✓	✓	✓	✓
Calculations involving fractions	1.4	✓	✓	✓	✓	✓	✓	✓	✓
Percentages, percentage difference	1.5	✓	✓	✓	✓	✓	✓	✓	✓
Indices and standard form	1.6	✓	✓	✓	✓	✓	✓	✓	✓
Ratio and proportion	1.7	✓	✓	✓	✓	✓	✓	✓	✓
M2 Handling data									
Approximating numbers (d.p. and s.f.)	2.1	✓	✓	✓	✓	✓	✓	✓	✓
Making estimates if the results of calculations	2.2	✓	✓	✓	✓	✓	✓	✓	✓
Finding the arithmetic mean and weighted mean	2.3	✓	✓	✓	✓	✓	✓	✓	✓
Units in chemistry	2.4	✓	✓	✓	✓	✓	✓	✓	✓
The ideal gas equation/converting volumes	2.5	✓	✓	✓	✓	✓	✓	✓	✓
M3 Algebra									
Symbols used in algebra ($=, <, \ll, \gg, >, \propto$)	3.1	✓	✓	✓	✓	✓	✓	✓	✓
Transposition of formulae	3.2	✓	✓	✓	✓	✓	✓	✓	✓
Representing a variable using more than one letter	3.3	✓	✓	✓	✓	✓	✓	✓	✓
Solving equations	3.4	✓	✓	✓	✓	✓	✓	✓	✓
Logarithms	3.5	✓	✓	✓	✓	✓	✓	✓	✓

Topic	Section	WJEC	AQA	OCR	Edexcel	CIE	CCEA	IB	Cam Pre-U
M4 Graphs									
Straight line graphs	4.1	✓	✓	✓	✓	✓	✓	✓	✓
Linear relationships	4.2	✓	✓	✓	✓	✓	✓	✓	✓
Drawing and using the slope of a tangent to determine rate of change	4.3	✓	✓	✓	✓	✓	✓	✓	✓
M5 Geometry									
2D shapes	5.1	✓	✓	✓	✓	✓	✓	✓	✓
3D shapes	5.2	✓	✓	✓	✓	✓	✓	✓	✓
Angles	5.4	✓	✓	✓	✓	✓	✓	✓	✓
Viewing molecules in 2D and 3D	5.5	✓	✓	✓	✓	✓	✓	✓	✓
C6 The amount of a substance									
Relative atomic mass	6.1	✓	✓	✓	✓	✓	✓	✓	✓
Definition of a mole	6.2	✓	✓	✓	✓	✓	✓	✓	✓
Relative molecular mass	6.3	✓	✓	✓	✓	✓	✓	✓	✓
Relative formula mass	6.4	✓	✓	✓	✓	✓	✓	✓	✓
Molar mass	6.5	✓	✓	✓	✓	✓	✓	✓	✓
Formula linking mass, molar mass and number of moles	6.6	✓	✓	✓	✓	✓	✓	✓	✓
Finding the percentage by mass of an element in a compound	6.7	✓	✓	✓	✓	✓	✓	✓	✓
Finding the number of particles in n moles of a substance	6.8	✓	✓	✓	✓	✓	✓	✓	✓
C7 Balancing and using chemical equations to predict masses									
Writing and balancing equations	7.1	✓	✓	✓	✓	✓	✓	✓	✓
Stoichiometry	7.2	✓	✓	✓	✓	✓	✓	✓	✓
When one of the reactants limits the reaction	7.3	✓	✓	✓	✓	✓	✓	✓	✓

Topic	Section	WJEC	AQA	OCR	Edexcel	CIE	CCEA	IB	Cam Pre-U
C8 Radioactive decay and half-life									
Radioactive decay	8.1	✓							
Half-life	8.2	✓							
C9 Atom economy and percentage yield									
Percentage yield	9.1	✓	✓	✓	✓		✓		✓
Percentage atom economy	9.2	✓	✓	✓	✓		✓		✓
C10 Determining relative atomic mass from percentage or relative abundance									
Relative abundance of isotopes	10.1	✓	✓	✓	✓	✓	✓	✓	
Percentage abundance	10.2	✓	✓	✓	✓	✓	✓	✓	
C11 Concentrations									
Converting volumes	11.1	✓	✓	✓	✓	✓	✓	✓	✓
Concentrations in parts per million	11.2	✓	✓	✓	✓	✓	✓	✓	✓
Concentrations used to calculate masses	11.3	✓	✓	✓	✓	✓	✓	✓	✓
C12 The ideal gas equation									
The ideal gas equation	12.1	✓	✓	✓	✓	✓		✓	✓
C13 Acid-base titrations									
Acid base titrations	13.1	✓	✓	✓	✓	✓	✓	✓	✓
C14 Determining empirical and molecular formulae									
The difference between empirical and molecular formulae	14.1	✓	✓	✓	✓	✓	✓	✓	✓
C15 Calculating the formula of a hydrated salt									
Finding the formula of a hydrated salt	15.1			✓			✓		

Topic	Section	WJEC	AQA	OCR	Edexcel	CIE	CCEA	IB	Cam Pre-U
C16 Enthalpy changes and calorimetry									
Enthalpy changes	16.1	✓	✓	✓	✓	✓	✓	✓	✓
Hess's law	16.2	✓	✓	✓	✓	✓	✓	✓	✓
Born Haber cycle	16.3	✓	✓	✓	✓	✓	✓	✓	✓
C17 Bond enthalpies									
Using the enthalpy change for a reaction to work out a bond enthalpy	17.1	✓	✓	✓	✓	✓	✓	✓	
C18 Rates of reaction									
Definition for rate of reaction	18.1	✓	✓	✓	✓	✓	✓	✓	✓
Examples in the use of the rate equation	18.2	✓	✓	✓	✓	✓	✓	✓	✓
Finding a rate of reaction usuing a concentration-time graph	18.3	✓	✓	✓	✓	✓	✓	✓	✓
Determining the units of the rate constant, k	18.4	✓	✓	✓	✓	✓	✓	✓	✓
Determining the rate equation using the initial rate method	18.5	✓	✓	✓	✓	✓	✓	✓	✓
Finding the rate determining step for a reaction	18.6	✓	✓	✓	✓	✓	✓	✓	✓
The Arrhenius equation	18.7		✓	✓	✓		✓		✓
C19 The shapes of simple molecules									
The shapes of simple molecules	19.1	✓	✓	✓	✓	✓	✓	✓	✓
Electron pair repulsion theory	19.2	✓	✓	✓	✓	✓	✓	✓	✓
Simple molecules where there are both bonding pairs and lone pairs	19.3	✓	✓	✓	✓	✓	✓	✓	✓
The shapes of ions	19.4	✓	✓	✓	✓	✓	✓	✓	✓
C20 The solubility of compounds in water									
Dilute solutions, saturated solutions and supersaturated solutions	20.1	✓							
C21 Oxidation states (or numbers)									
Oxidation and reduction	21.1	✓	✓	✓	✓	✓	✓	✓	✓
Oxidation states (also called oxidation numbers)	21.2	✓	✓	✓	✓	✓	✓	✓	✓
Determining whether a reaction has been oxidised or reduced	21.3	✓	✓	✓	✓	✓	✓	✓	✓

Topic	Section	WJEC	AQA	OCR	Edexcel	CIE	CCEA	IB	Cam Pre-U
C22 Redox reactions									
Redox reactions	22.1	✓	✓	✓	✓	✓	✓	✓	✓
Proving that a reaction is a redox reaction	22.2	✓	✓	✓	✓	✓	✓	✓	✓
Proving that a reaction is not a redox reaction	22.3	✓	✓	✓	✓	✓	✓	✓	✓
C23 Electrochemical cells									
Electrodes and half-cells	23.1	✓	✓	✓	✓	✓	✓	✓	✓
Using standard electrode potential data	23.2	✓	✓	✓	✓	✓	✓	✓	✓
The Nernst equation	23.3					✓			
Electric charge and the Faraday constant	23.4					✓			
C24 Entropy and free energy change									
Entropy	24.1	✓	✓	✓	✓		✓	✓	✓
Gibbs free energy	24.2	✓	✓	✓			✓	✓	✓
C25 Equilibria									
Reactions in equilibrium	25.1	✓	✓	✓	✓	✓	✓	✓	✓
Finding the concentrations at equilibrium	25.2	✓	✓	✓	✓	✓	✓	✓	✓
Using partial pressures rather than concentrations	25.3	✓			✓	✓	✓		✓
C26 Calculations involving pH									
pH and the pH scale	26.1	✓	✓	✓	✓	✓	✓	✓	✓
C27 Analytical techniques									
Mass spectroscopy	27.1	✓	✓	✓	✓	✓	✓	✓	✓
Infrared spectroscopy	27.2	✓	✓	✓	✓	✓	✓	✓	✓
nmr spectroscopy	27.3	✓	✓	✓	✓	✓	✓	✓	✓
The relationship between energy and frequency	27.4		✓				✓		

Index